POWER
STEERING

THE COMPLETE BOOK OF FARMALL TRACTORS

EVERY MODEL 1923–1973

TEXT BY
ROBERT N. PRIPPS

PHOTOGRAPHY BY
ANDREW MORLAND

motorbooks

F-30
McCORMICK-DEERING
FARMALL

CONTENTS

INTRODUCTION

A Short History of the Tractor to 1924

To the farmer, the invention of the tractor was almost as important as the invention of the reaper. It was Cyrus Hall McCormick's reaper, introduced in 1831, that initiated the power farming era, when mechanisms became a substitute for physical strength and an individual could accomplish more in a day than several men could have previously. The tractor, however, did not suddenly burst on scene as the reaper did. As they used to say about things that took time to develop, the tractor was "slow-a-bornin'."

The story of the tractor properly begins with the development of steam power. Although Scottish inventor James Watt is credited with the eighteenth-century inventions that improved the steam engine, it was almost a hundred years later, in 1849, that A. M. Archambault & Company of Philadelphia made the first farm steam engine. When newspaper man Horace Greeley saw that engine at work he wrote, "The time must be at

(opposite)
The amount of human and animal labor that went into producing a loaf of bread before the turn of the century is mind-boggling. First, the soil had to be tilled by hand, or with an animal-powered plow. Next, the seed was sown by hand. Farmers harvested the wheat with a sickle or scythe and then threshed it with a flail, or by leading sharp-footed animals over it as it lay on the "threshing floor." The wheat then had to be winnowed by hand to separate the wheat from the chaff. When Cyrus Hall McCormick invented the McCormick reaper in 1831, all of this changed. The 1857 McCormick Patent Reaping and Mowing Machine shown here, manufactured from 1852 through 1865, cut harvesting times from one day per acre to one hour per acre. *Photo courtesy of Washington Historical Society*

Waterloo Boy advertisement from a 1917 *Farm Implements* magazine.

hand when every thrifty farmer will have such an engine of his own, and chopping straw, turning grindstone, cutting wood, churning, threshing, etc., will cease to be a manual and become a mechanical operation."

The first steam engines were stationary or portable, but the wheels were not powered. Later, manufacturers developed the traction engine, a machine with powered wheels that drivers could steer from a cab or platform. The steam traction engine was essential in driving the threshing machine. In the western states and on the vast Canadian prairies, giant steamers also pulled plow gangs so large that they boggle the mind even today. But the steam traction engine proved to be too unwieldly, costly, and cumbersome for most farmers. A more practical farm tractor was unavailable until German inventor Nicholas Otto patented the Otto internal combustion engine in 1876. The Otto engine featured the classic four-stroke principle still in use today. When Otto's patents expired in 1890, companies all over the world jumped into the engine business. Before long, there were more than 100 brands of stationary Otto-type engines on the market.

Gasoline tractors made their debut in the 1890s. For the most part, they were steam traction engines chassis fitted with gasoline engines. Most notable of these hybrids was the 1892 Froelich, created when John Froelich of Froelich, Iowa, mounted a Van Duzen engine on a Robinson chassis. His machine, rigged with a reversing gear, clutch, and steering mechanism, was the first gasoline tractor that could propel itself both forwards and backwards, allowing farmers to tow a threshing machine along roads. When operators reached the field, they could engage a flat belt to power the thresher. Unfortunately, John Froelich seemed to be the only one capable

— WILLIAM DEERING —

William Deering was one who capitalized when McCormick delayed patenting his new harvesting system developments. By the time he was forty years old, Deering had made a substantial fortune in dry goods in the East. Deering traveled to Chicago in 1870 to investigate a land investment and happened to call on an acquaintance, Elijah Gammon. Elijah had invested in the newly invented and promising Marsh Harvester and offered Deering the opportunity to get in on it, too. Deering did, and in two years his investment had doubled. Deering then moved to Chicago and became active in the farm implement business.

— CYRUS HALL McCORMICK —

Cyrus Hall McCormick was born to Robert and Mary Ann McCormick on February 15, 1809. The McCormicks were fairly wealthy Virginia farmers with more than 500 acres of land. Robert was a man of mechanical ingenuity. He had tinkered with a mechanical reaper for many years without much success. When son Cyrus was old enough, he most likely joined his father in reaper experiments. When Cyrus grew up, he began doing these experiments on his own, in an effort to lighten the labor-intensive job of grain harvesting on the McCormick farm. In July 1831, Cyrus McCormick demonstrated a successful grain-reaping machine. Two men and one horse could harvest about one acre per hour. The accepted rate for a man with a cradle scythe was an acre per day.

Besides selling a few reapers to neighbors, C. H. McCormick was interested in running his own farm and in refining other farm inventions that he and his father had developed. Meanwhile, Obed Hussey patented a reaping machine in 1833.

Cyrus then secured his own patent and battled Hussey in the courts and on the fields until Hussey's untimely death in a railroad accident in 1860. Unfortunately for both, their patents ran out at the same time that mechanical reaping really caught on. Other farm implement companies brought out reapers of their own and then began competing for the rights to new inventions, such as the self-raking reaper, the binding reaper, and the twine knotter.

Cyrus Hall McCormick

Like John Deere and J. I. Case, McCormick soon recognized that the market was in the Midwest, so he moved his production enterprise, the McCormick Harvesting Machine Company, to Chicago, Illinois, in 1847. The business was incorporated in 1879.

Cyrus Hall McCormick died in 1884, but his widow and his son, Cyrus Jr., obtained controlling interest in the corporation.

of operating the Froelich tractor effectively. Nevertheless, the Froelich was the forerunner of the Waterloo Boy line, which John Deere eventually acquired.

Two young engineers named Charles Hart and Charles Parr created the next gasoline tractor of note. The enterprising pair started making and selling stationary engines as a way to fund their engineering studies at the University of Wisconsin. They later opened a tractor factory in Charles City, Iowa, and in 1901, they built their first tractor. After thoroughly testing it, Hart and Parr sold it to a farmer in 1902. They improved upon the design and sold fifteen of the new variety the following year. (Amazingly, half of these first Hart-Parrs were still in use seventeen years later.) At the time, Hart and Parr's

Two Minnesota farmers pause for a photograph alongside their 1912 Mogul. ***Photo courtesy of Minnesota Historical Society***

factory was the only business in America devoted exclusively to tractor manufacturing. It was, in fact, a Hart-Parr employee that coined the word "tractor," which was short for "traction engines."

At the beginning of the twentieth century, Canada was the leading tractor market. Preparing the country's wide-open spaces for farming was a daunting task. In fact, beginning in 1908 and continuing for several years, Canadians hosted great plowing contests near Winnipeg as a way of enlisting others to help clear the land. Huge machines with as much as 50 horsepower made quick work of opening the vast western prairies. While the development of these behemoths certainly advanced the industry, the average farmer in the United States did not need machines of this magnitude.

By 1910, International Harvester climbed ahead of Hart-Parr as the leading producer of these big tractors. In 1912, IH, Rumely, and Hart-Parr produced almost 10,000 tractors, many of which fell into the 20,000-lb. class. With the partial crop failures and economic downturn of 1914, the land boom in western Canada collapsed, dragging the market for these huge tractors down with it. Industry leaders soon realized that if they were to have a future in the tractor industry, they would have to service and produce smaller machines.

In 1913, Bull Tractor Company of Minneapolis produced the first tractor for the small farmer. This remarkably different kind of a tractor was driven by a single "bull" wheel, making an expensive differential unnecessary. An unpowered "idler" wheel, located on a crank axle so that the tractor could run level with the bull wheel in the plow furrow, provided balance. The Bull had a two-cylinder, horizontally opposed engine of 12 horsepower. Instead of a transmission gearbox, a cone clutch connected the engine to the bull wheel. The little tractor only weighed 3,000 lb. and sold for the unbelievably low price of $395. The price alone enticed many an American farmer to buy a Bull, although it wasn't long before owners were complaining that the new tractor was disappointing them at every turn. Nevertheless, for several years, the Bull even outsold International Harvester's line of much larger tractors.

The sales success of the Bull (followed quickly by an improved "Big Bull") prompted other manufacturers to get in line with the smaller but sounder tractors. IH introduced

The 1916 Big Bull tractor, as it appeared in a *Farm Implements* ad.

NEW DAY

their Titan 10-20 and Mogul 8-16 machines. Case and Allis-Chalmers joined the field. Hart-Parr offered its one-wheel-drive Little Devil with a two-cycle engine. Huge tractors were still much in vogue, however, with Twin City offering a 90-horsepower machine with 84-inch drive wheels.

When the United States entered World War I in 1917, demand for tractors escalated as farm lads joined the military, leaving tractors to make up for the lost labor. Tractor production doubled from the previous year with almost 63,000 being built, some 15,000 of which were exported to Allied countries. International Harvester led the way in production, followed by Case, Avery, and Moline.

With around 200 companies making tractors, production doubled again in 1918. This was the same year that auto magnate Henry Ford entered the field with his Fordson. Cyrus McCormick III, who was the head of International Harvester at that time, writes about Ford's affect on the small tractor business in his 1931 book *The Century of the Reaper*:

The harvester war of the Eighteen-Nineties was cruel, disastrous to the weaker combatants, and yet it was inspiring in the way its testing brought out the finer qualities of men. But in the first twenty International years, competition had perhaps become routine. Henry Ford's presence in the implement province and the new type of competition he soon introduced returned the industry for a time to the atmosphere of battle. War hurts, but its searing occasionally furnishes an urge to continued growth.

In 1918, when he sold his first tractors in the United States and Canada, Ford distributed them through government agencies as a war measure. To have convinced worried statesmen and the public that the tractor was a new device twelve years after many tractor builders had attained large production, was a supreme feat of salesmanship. To win the news columns of the metropolitan press for a discussion of how the magic name of Ford had, at a stroke, provided an answer to a supposedly unrecognized demand for farm power, was magnificent advertising. Whatever Ford did was deemed to be for the real benefit of the public; and wherever he led,

— INTERNATIONAL HARVESTER COMPANY —

Competition in the farming industry was fierce, but none of the competitors were making much money. Talk of amalgamation of McCormick's and Deering's firms had begun even before Cyrus Sr. died in 1884. In the fall of 1890, McCormick, Deering, and eighteen other lesser competitors attempted to consolidate under the banner of the American Harvester Company. When the smaller companies established their own unrealistically high valuation in order to get a bigger piece of the whole, bankers saw through the scheme and refused to loan working capital to the new corporation. McCormick and Deering knew they would be foolish to support the weaker constituents, and they too balked. With that, the American Harvester Company foundered.

A severe recession in 1896 and 1897 prompted the two harvesting dynasties to again think seriously about joining forces. George Perkins, a partner in the financial firm of J. P. Morgan, pushed the merger efforts, which included acquiring three smaller firms. Perkins proposed a ten-year trust that would hold all the stock of International Harvester, with Perkins, McCormick, and Charles Deering (William's son) as trustees. So, on July 28, 1902, International Harvester was born. Perkins picked the name to reflect what he saw to be a global enterprise. Harvester had an 85 percent share of U.S. sales of harvesting machines. Besides the basic manufacturing facilities, Harvester had iron foundries, twine factories, sawmills and timberlands, hemp farms, coal and iron mines, and its own railroad.

Perkins rode herd on the "millionaire officers" and appointed a professional general manager to run day-to-day operations. The first was a capable man named Clarence Funk. After ten years, however, the McCormicks borrowed $5 million from J. D. Rockefeller and gained control of the corporation. Their first move was to replace Funk with a GM of their own choosing, one Alexander Legge. It was Legge's vision and fortitude that would give birth to the Farmall tractor.

International Harvester, also known as "Harvester" or simply "IH," went on to become the largest, most successful truck, tractor, and agricultural implement company in the world. Some of the very best multipurpose tractors came from IH under the trade name Farmall. But it was seldom smooth sailing for the Chicago-based International Harvester. Competition, economic depression, labor problems, drought in the farm belt, and difficulties obtaining working capital plagued Harvester and the rest.

consumers followed. It is therefore not surprising that his tractor business increased by leaps and bounds.

During 1918 there were 133,000 tractors made in the United States. Ford had already usurped Harvester's leadership, International was second, and Case third. The expected drop in the wartime demand did not eventuate, and production mounted rapidly to the astonishing total of 203,000 machines in 1920. By this time Ford was far in the lead and was making several times as many as Harvester. For the next year or two, three-quarters of all tractors made were Fordsons. During the period of depression, sales fell off to a quarter, inventories of materials, and unsold tractors were huge and high-priced, and the prospects were gloomy. Then, early in 1922, Henry Ford cut the price of tractors.

That February morning is another of the many business hours I treasure in my memory. I had taken Mr. Legge, the Company's beloved and hard-boiled general manager, on a visit to the new motor-truck installation at Springfield Works. As we were arguing some

1919 International 8-16

The 8-16, forerunner of the famous Farmall, proved there was a market for a powerful, lightweight tractor. Built between 1917 and 1922, the 8-16 was sometimes called the International Junior or the Mogul Junior. The design was based on the Model G International truck, featuring the same engine (with the radiator behind), downward-sloping hood, and a three-speed transmission. The 8-16's overhead-valve engine with a bore and stroke of 4x5 inches produced 16 belt horsepower and 8 horsepower on the drawbar at 1,000 rpm.

problem which then seemed important, the telephone rang—Chicago wished to speak to Mr. Legge. We could, of course, hear only his side of the conversation. There was much talk from the other end, and then an explosion from Alex: "What? What's that? How much? Two hundred and thirty dollars? Well I'll be . . .What'll we do about it. Do? Why, damn it all—meet him, of course! We're going to stay in the tractor business. Yes, cut two hundred and thirty dollars. Both models—yes, both. And say, listen, make it good! We'll throw in a plow as well!"

By the time of this phone conversation, Ford's production rate was filling his lots with unsold tractors. He had to move them! And, rather than lose its place in a business it had pioneered, Harvester rose to meet Ford's challenge by also selling at less than cost. The winner, of course, was the farmer. With the Fordson selling for as low as $395,

The 1915 Titan 10-20, as it appeared in a 1967 International Harvester calendar.

The International 8-16 in use, circa 1918.

power farming was now within the reach of all but the smallest farmer. Harvester countered with their trusty Titan 10-20 and their newer International 8-16 (based loosely on an International motor truck). But even with Legge's $230 price cut, the IH tractors still cost as much as the Fordson.

Following tactics learned in the Great Harvester Wars of the 1890s, when a Harvester salesman learned of a pending Fordson sale, he challenged the farmer and the Fordson dealer to a competition. Each tractor would pull a two-bottom plow with 14-inch bottoms through whatever conditions the farmer chose. If the International tractor outperformed the Fordson, the farmer could change his mind about the purchase and decide to buy the IH tractor instead. Although the IH cost almost twice as much as the Fordson, the deal included a free plow. In many conditions, the lightweight Fordson would demonstrate a marked lack of traction. In other conditions, the Fordson's performance was not all that bad, and the IH tractors ended up looking somewhat dated and obsolete next to the carlike Fordson.

Competition with the Fordson, however, prompted International Harvester to rapidly adopt automotive production methods and standardize its parts. It also instigated the further development of the power take-off (PTO) and the application of industry standards for things like spline sizes and belt speeds. The PTO gave the tractor more versatility around the farm. The idea of an all-purpose tractor wasn't entirely new. The introduction of the Moline Universal in 1917 is a case in point. But the giant step in power farming occurred when International Harvester introduced the McCormick-Deering Farmall in 1924.

— THE GREAT HARVESTER WAR —

In his book *The Century of the Reaper*, Cyrus McCormick III recounts a commercial fight that took place in the late 1880s near Pomeroy, Iowa. The tale comes from the report of the McCormick general agent.

"Our dealer telegraphed me that Champion had pulled into the same field where he had sold an eight-foot McCormick binder and were trying to break up the sale.

"When I got there, the farmer was having trouble with our machine, and the Champion boys were giving him plenty of poison about it and me. I got the machine fixed so that it worked properly. But Champion had made the famer some kind of a special price; also, they had notified every purchaser of a McCormick binder in the neighborhood to come and see this binder fail, with a view of getting them to cancel their orders.

"A big crowd of machine men had come to the hotel that night and the Deering fellows said they would come in too and show us both up.

"I got up at three o'clock in the morning and drove out to the farm and woke the farmer up. He put on his pants and came out to the barn where I gave him such a sales talk that he was absolutely convinced the McCormick was the best machine.

"In the morning there were at least 150 farmers there. The Deering outfit was the first to start. They had a new machine all decorated with flats and four big gray horses. But when the first bundle of tangled barley came through it choked and they were done.

"The farmer, driving our binder, was having no trouble, but I caught a Champion man trying to put a handful of straw in our elevator chains to foul them. I grabbed him by the neck and he fell down in the stubble. Then the whole Champion crew started after me, but somebody got between us. They started to abuse the farmer, a big powerful man, and he struck the Champion dealer. The farmer's old father stopped the fight, but the whole competition broke up into a row. Finally, Champion left in disgrace without having driven us from the field."

1927 Farmall Regular.
Ralph W. Sanders/Motorbooks Archive

CHAPTER 1

The Original Farmall 1924–1932

"However, his (Legge's) best answer to Ford can be expressed in one word: 'FARMALL'"
—Barbara Marsh, author of _A Corporate Tragedy_

When the world saw the new Farmall, the age of the Fordson was over, at least in America. Yes, Fordson production in Dearborn continued into 1928, and yes, Ford sold a bunch of Fordsons in the intervening years, but in 1924, International Harvester began turning a profit on their lines and, in about 1926, regained the lead from Ford in agricultural tractor production. The key to Harvester's recovery was the revolutionary Farmall.

Prior to 1924, tractors were tailored for plowing and belt work only, leaving the rest of the daily tasks around the farm to the farmer and horses. Several implement makers had looked into the possibility of engine-driven mowers, rakes, and cultivators, but nothing had become of their experiments. It took the onslaught of the Fordson to inspire the engineers at International Harvester to dust off their designs and get serious about an all-purpose tractor. And International Harvester's experimental department delivered.

The creativity, loyalty, and gumption of the company's founder, Cyrus Hall McCormick, must have rubbed off on the Harvester engineers, or the "McCormick Men," as they were called. Alexander Legge, company president, was but one of the great McCormick men. The experimental department also included department manager Edward Johnston and colleagues John Anthony, C. W. Mott, and P. H. Danly. Johnston had talked to Legge about their ideas for a more versatile farm tractor as early as 1910.

McCormick-Deering catalog, circa 1930s.

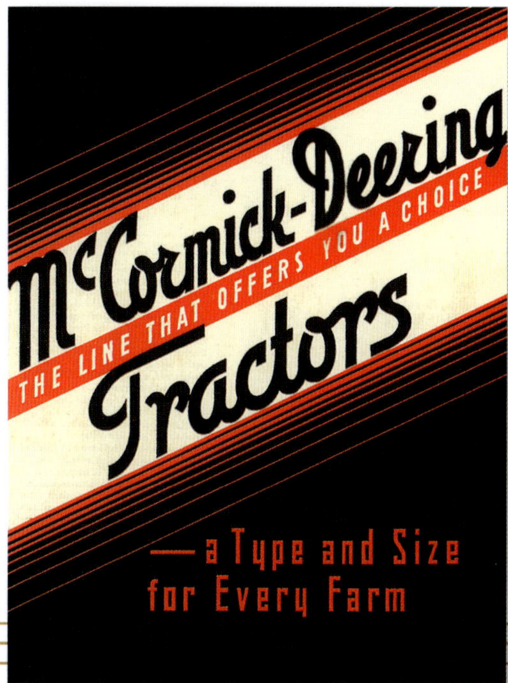

Ed Johnston, a pioneer in International Harvester engineering.

In fact, they went so far as to file for a patent on a machine they called a motor cultivator. The influential Bert R. Benjamin, then head of the McCormick Works in Chicago where the great McCormick implements were assembled, also backed the motor cultivator idea.

The idea for a multipurpose power-farming machine goes back even farther than 1910, however. When International Harvester formed in 1902, Deering engineer John F. Steward brought ideas for a motor-driven harvesting machine to the new company. The fairly lightweight machine would be capable of pushing or pulling implements such as corn huskers, strippers, or binders and operating with an underslung plow. The Steward "tractor" was the first to use close-set wheels that pivoted for steering.

Ed Johnston was an old-timer with the company. He had started at age fifteen during the Harvester War of the 1890s and had been instrumental in the improvement of many of McCormick's products. He had patents covering mowers, binders, knotters, and other machines. He even made himself a vehicle he dubbed an "Auto Buggy," which he used to get to and from work. Cyrus McCormick III took a liking to

(right and opposite) Experimental versions of the all-purpose Farmall tractor, circa 1920.

the buggy and commissioned Johnston to create one in the spirit of a farm wagon. The vehicle became the forerunner of the International truck.

Between 1915 and 1921, the experimental department made several versions of the machine they called the motor cultivator, always keeping Alex Legge apprised of their progress or difficulties. The machine emerged with its engine in back, directly over the narrow-set drive wheels. This whole assembly could be rotated for steering. The operator sat in the middle with the widespread wheels and the cultivator apparatus in front. At that time the machine had two modes of steering. Foot pedals angled the widespread front wheels, making small corrections to prevent the cultivator shovels from hitting the crop. At the ends of the rows, a driver could use the steering wheel to unlock and rotate the engine and narrow-set back wheels. This allowed short turns in order to get into the next rows. Two main problems persisted: The motor cultivator was too expensive, and the engine mounting caused the center of gravity to be too high, making the machine unstable on side hills. Nevertheless, the motor cultivator went into production, and some 350 examples were sold between 1917 and 1921.

One of the farmer's main jobs was the cultivation of tall crops, especially corn. Implement makers attempted to satisfy the demand with a variety of engine-driven cultivators. In 1919, after a few years of development, five companies, including International Harvester, participated in the first Corn Belt cultivator demonstration in Blue Mound, Illinois. It soon became apparent to all, however, that the engine-driven cultivator was too expensive for the average farmer, who could simply not afford to have such an investment sit idle for eight or nine months of the year. This led most, including Harvester, to abandon development of a cultivator-only machine and work to develop the all-purpose tractor instead.

In 1920, the International Harvester Motor Cultivator became known as the "Farmall" in internal IH documents. The last written mention of the Motor Cultivator was in August 1919. It was at about this time that the engine was relocated ahead of the now-single rear wheel, and the widespread front wheels became the drivers. The Farmall, which was actually configured to be reversible, was tested as a pushing or pulling machine. The driver could swing the seat around the steering wheel to face the opposite direction. The drive wheels could, therefore, be either in the front or the back, steered by the single or dual-narrow wheels on the opposite end. Up until this time, the engine had been mounted crosswise in order to avoid expensive bevel gears. In late 1920, however, an experimental version featured an inline engine. This alteration

First production Farmall, 1924.

lowered the center of gravity and dramatically improved the driver's visibility when operating in what would become the conventional direction of travel.

But without a push from management, plans for the all-purpose Farmall tractor languished until July 1921. Then, at a meeting of top executives, Alexander Legge outlined the takeover of the tractor market by Henry Ford. Legge called for a report from the experimental department. Edward Johnston quickly reported on the status but said lack of management support had slowed progress in the development of attachments. Johnston insisted, however, that the Farmall could outperform the Fordson in every way.

1924 Farmall Regular
Before 1924, tractors were designed for either plowing or for driving implements via the flat belt pulley. The Farmall performed all of those tasks and cultivated crops. The farmer could also use the Farmall's rear power takeoff (PTO) to drive harvesters, pumps, and generators by shaft. With its close-set front wheels and large-diameter rears, the Farmall established the "tricycle" configuration tractor. This unique design influenced the industry for the next fifty years.

In that same year, Benjamin's McCormick Works mounted a cultivator near the single front wheel. Now, with the driver's seat located by the wide-spaced drive wheels (now in the rear), operators could steer the single "front" wheel to accurately control the cultivator shovels. The system worked so well that the thought of developing a reversible tractor died a painless death. Benjamin also revised the basic configuration of the tractor to reduce its size and weight. The experimental team continued to work on the Farmall, improving the tractor's appearance and ease of manufacture. Although the name "Farmall" had been used internally since 1919, it was not registered as a trademark until 1923. The company's naming committee still did not officially recognize it until February 5, 1924.

IH built twenty-two Farmall tractor prototypes in 1923. Engineers conducted field tests of the machines on actual farms, improving parts that did not hold up. Meanwhile,

1924 Farmall Regular
Harry Lee, of Elnora, Indiana, a well-known and respected IH collector and restorer, stands proudly beside his Farmall Regular. He maintained that at least some of the 1924 Farmalls were painted silver at the factory. Other restorers who have stripped the old paint layers from their 1924 Regulars confirm that silver was sometimes the original finish.

internal opposition to the Farmall began to develop by those fearing that if the Farmall were put on the market it would compete with the McCormick-Deering Model 10-20 rather than the Fordson. Salespeople had finally begun to hold their own against the low-cost Fordson, and they didn't want anything to sabotage the progress they'd made. Nevertheless, in 1924 Legge gave the go-ahead to build 200 Farmalls to sell for a loss-leader selling price of $825. Naturally, volume production would bring the costs down, but these tractors were practically hand-built. Salesmen and engineers kept close tabs on the field performance. Although the first of these tractors went to an Iowa farmer, most went to Texas, where IH dealers had had a more difficult time getting the 10-20 standard-tread tractor to ranchers in remote areas.

For the 1925 production run, engineers made more improvements. They corrected chronic difficulties with the differential housing and added an air cleaner to the intake pipe, replacing the flannel cloth that had been used previously. By this time, IH was able to offer a variety of implements, from a grain binder with shocking attachment to a cord wood saw. In September 1925, the Farmall was tested and accepted by the University

The 1924 Farmall All-Purpose tractor, as featured in a 1967 International Harvester calendar.

of Nebraska, a prerequisite to selling a tractor anywhere in the country. During approximately thirty-nine hours of exacting tests, the Farmall performed without difficulty, requiring no repairs or adjustments. It demonstrated a maximum drawbar horsepower of 12.7 and a maximum belt rating of 20.05 during Test No. 117.

The Farmall was equipped with a 221-ci vertical four-cylinder overhead-valve (OHV) engine of Harvester's own design and manufacture. Bore and stroke were 3.75x5.0 inches and the rated operating speed was 1,200 rpm. The tractor had a three-speed transmission.

By 1926, the Rock Island Farmall plant was in production. The price of the Farmall was raised to $950, $100 more than the 10-20, to show that the Farmall had more utility.

1924 Farmall Regular
This Regular was the eighty-seventh Farmall made for sale to the public. Characteristics of the 1924 Farmall include a seat spring-bolted to the cast-iron differential housing, brakes with external return springs, governor and spark controls located near the operator's right foot, front wheel hubs with flat roller bearings, and an erect exhaust pipe with a cloth sock air cleaner. The serial number is stamped on the block rather than on the frame.
Photo courtesy of John Wagner

1928 Farmall Regular
This original decal had survived seventy-five years of wind, rain, and sunshine when this photograph was made.

Legge authorized a production level of 25 Farmalls per day, although 4,430 were actually built that year.

A "Fairway" version, identical to the original, except equipped with wide, flat steel wheels, was also produced. A total of 9,902 Farmalls rolled off the assembly line in 1927, and by 1930, production reached 200 per day. By April 1930, IH celebrated the sale of 100,000 Farmalls since 1925. Total production of the original (or regular) Farmall through 1932 totaled 134,647. In the previous five years, Harvester also sold 200,000 of its Model 10-20 standard. And, from 1921 through 1930, IH sold some 150,000 of its expensive ($1,500) Model 15-30 standard-tread tractor. Business was booming.

1927 Farmall advertisement.

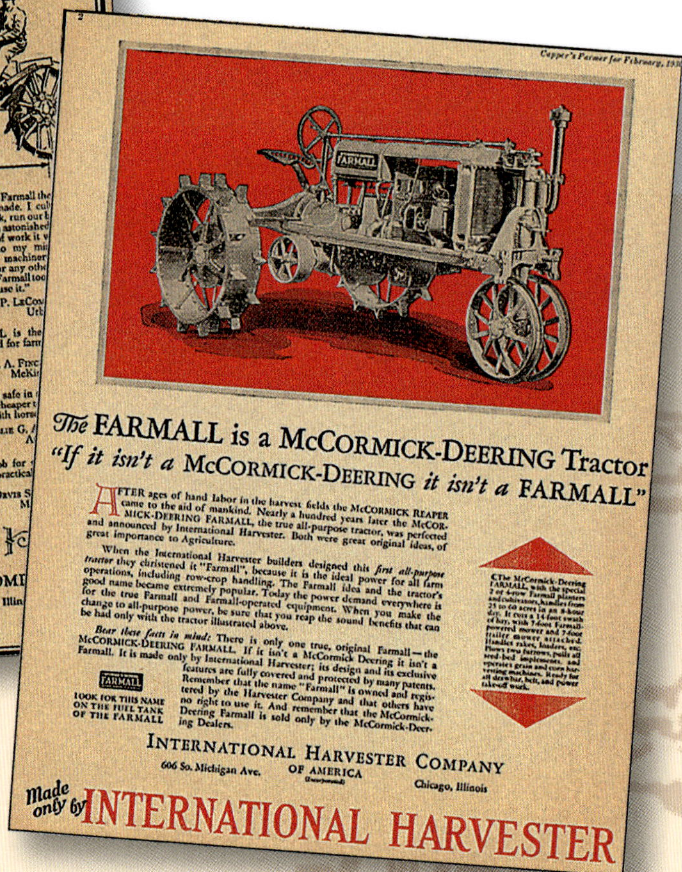

Farmall ad from the February 1930 issue of *Capper's Farmer.*

1930 Farmall Regular

The first Farmall grew out of motor cultivator projects at International Harvester. It was at first called "Farmall" until siblings Fairway and the Farmall F-30 came along. The unofficial moniker "Regular" was tacked on to the original tractor's name for differentiation. The Farmall Regular was powered by a four-cylinder overhead-valve engine with a bore of 3.75 inches and a stroke of 5.0 inches. This nicely restored Regular features rubber tires, although originally the tractor had steel wheels. International Harvester offered rubber tires for the first time in 1934 on the Farmall F-20.

1931 Farmall Regular
From the time the Farmall was introduced in 1924 until the end of its production run in 1934, the tractor's design changed very little. *Ralph W. Sanders/ Motorbooks Archive*

Meanwhile, Fordson moved all production from Detroit to Cork, Ireland, in 1928. In the preceding ten years, almost 850,000 of the tractors had been built in the U.S. factories and in the Cork, Ireland, factory. Strangely, Ford had virtually no production of any kind in 1928. The last of the Model T cars came out in 1927. The first Model A was released in 1929. The few Fordsons produced in 1928 (about 8,000) were built from parts produced in 1927. Ford reasoned that the Model A would be such a success that the Fordson's space would be needed for car production. Also, Lord Percival Perry, Ford's friend and the head of British Ford, wanted Fordson production. With things going badly in the European car market, even the Fordson spares and parts business would be worthwhile. Fordson, however, was not out of the picture, as about 1,500 per month of an improved model were soon being imported. But the Fordson was never again a serious challenge to IH.

The Farmall tractor changed the basic configuration of the American farm tractor to that of the narrow-set front wheels and high-wide rear wheels—a configuration that would become known as the all-purpose, or general-purpose, or row-crop tractor.

In this 1929 advertisement, International Harvester recruits some of its best customers to sing praises of Farmall tractors.

COMPARISONS

Make/Model	CID	Cyl.	Belt/PTO HP	Drawbar Pull (lb.)	Fuel Cons.*	Weight (lb.)
Farmall Regular	**221**	**4**	**20.05**	**2,727**	**9.39**	**4,100**
Fordson F	251	4	22.28	2,142	8.95	3,175

*horsepower hours per gallon
Weight in pounds is as tested including ballast (if any) and driver.

Siblings demonstrate the capabilities of the all-purpose Farmall.

The "Master of Mechanical Power" ad appeared in the October 1930 issue of the *Country Gentleman.*

1935 Farmall F-12. The tractor shown has the steering-wheel lift mod of the F-14.
Ralph W. Sanders/Motorbooks Archive

CHAPTER 2

The Farmall Becomes a Series 1932–1939

***"Harvester is, of course, the greatest single agricultural enterprise in the world."* —*Fortune* magazine, 1933**

By 1931, Harvester general manager Alexander Legge had guided the Harvester ship through the shoals of government antitrust actions on one hand and below-cost competitive pricing by auto magnate Henry Ford on the other. The company was now on top as an agricultural entity. The federal government had given up on their efforts to break up the company, and Ford's Fordson had moved to Great Britain. Harvester's Farmall all-purpose tractor, which debuted in 1924, was exceeding all expectations, both in the company and on the farm. Harvester's other farm equipment and the company's truck line were also selling well. Further, a Republican was in the White House, and that suited the conservative Mr. Legge.

Edward Johnston was not for resting on laurels, however. He warned Harvester's executive council that Ford wasn't the company's only competition. John Deere had introduced its GP (general-purpose) tractor in 1929. It was off to a slow start, but Deere was known to vary and adjust models as necessary to meet almost individual needs. In 1930, Oliver came out with a very attractive tricycle row-crop tractor. Case also offered the neat Model CC row-crop machine, and Massey-Harris, the Canadian giant with tractor production facilities in Racine, Wisconsin, produced a revolutionary four-wheel-drive General Purpose. Johnston convinced the council to adopt an Increased Power Program in January 1930. The

1930s Farmall catalog cover.

This painting of a Farmall F-20 was featured in a 1971 International Harvester calendar.

The "Nothing but Praise for the Farmall" ad appeared in the November 1932 issue of the *Country Gentleman.*

executive council approved three sizes of new Farmall tractors, all with more power than the original.

In addition to the base model of this new line, called the Increased-Power Farmall, IH planned a larger intermediate Farmall with an upgraded 10-20 engine and an even bigger model called the Large Farmall with an upgraded 15-30 engine. Prototypes were built with some new features, including the cambered dual narrow front wheels, an optional wide front end, and a four-speed transmission.

Early in 1931, the IH naming committee named the base, or Increased-Power Farmall, the F-20 and gave it a two-plow rating. They named the Intermediate Farmall the F-30 and gave it a three-plow rating. The Large Farmall, called the F-40, received a four-plow rating. To differentiate the original Farmall from newer versions, the original unofficially became known as the "Farmall Regular." By the end of February 1931, the executive council's new works committee decided not to develop the F-20 or the F-40 any further at that time. They called for an additional ten prototype F-30s and for some test improvements to the Regular Farmall. A month later, they reinstated the F-20.

Farmers were serious about their Farmall F-20 tractors. ***Library of Congress***

McCORMICK-DEERING FARMALL 30—
The Big-Time Member of the Famous Farmall Family

THE Farmall 30 is the big-farm member of the Farmall family. It was added to the Farmall line five years ago to meet the needs of row-crop farmers in the 300-acre class or thereabouts, also to give the extra power required on many medium-size farms where hills or stubborn soil rule out a smaller tractor. The Farmall 30 extends the benefits of the Farmall system of farming to the large operator as the Farmall 12 brings those benefits within reach of the small operator.

The Farmall 30, like the Farmall 20, has four forward speeds. But with the added power of its 4¼ x 5 engine it will go ahead and handle 4-row planters and cultivators or pull three plows under severe working conditions with a speed and efficiency impossible in a smaller tractor. It is this ever ready all-day-and-night reserve of power, combined with Farmall's mastery of every row-crop farming operation, that has made the Farmall 30 a favorite wherever heavy seasonable demands must be handled on a strict time schedule. In backward seasons the Farmall 30 owner does not worry. He knows his tractor will perform at top speed 24 hours a day if necessary and that the amount of work he does is determined by his own wishes, not the capacity or endurance of his power plant.

The Farmall 30 and the various machines designed for it are protected by the same patents which cover the original Farmall. The Farmall 30, like the two smaller Farmalls, is delivered completely equipped with belt pulley and power take-off, ready for any and all types of machines and work. Combination manifold for either gasoline, kerosene, or No. 1 distillate.

WHAT THE F-30 WILL DO:

1. Will go onto any row-crop farm in the 200- to 300-acre class and take over virtually every power operation, eliminating any necessity for keeping horses or mules. 2. Will pull three 14-inch bottoms in most soils and plow from 9 to 13 acres a day; double-disk 30 to 40 acres; plant (4-row) 40 to 50 acres of corn or cotton; cultivate 40 to 60 acres; cut or pick (2-row) 14 to 18 acres of corn; operate a 2-row potato digger; harrow (peg tooth) easily 100 acres; drive a 28-inch thresher in average conditions and the larger sizes of ensilage cutters, huskers and shredders, hammer mills, and other belt machines. 3. Will do the work of 10 to 14 horses or mules, and do it at a substantial saving. 4. Will do a full day's work on from 25 to 30 gallons of fuel.

MAIL THE POST CARD ON PAGE 31 FOR MORE INFORMATION

FULL SPECIFICATIONS ON PAGES 16 AND 17

PAGE 10

This page from the McCormick-Deering "Tractor Power" catalog details the advantages of owning the Farmall F-30.

The Farmall F-30

The Farmall F-30 went into production in late 1931. Weighing in at 5,300 lb., the F-30 was almost three-fourths of a ton heavier than the Regular and about two feet longer. The four-cylinder OHV engine was similar to the 10-20 engine. Both the F-30 and the 10-20 engines displaced 284 cubic inches and had the same bore and stroke of 4.25x5.00 inches. The University of Nebraska tested the F-30 in October 1931. The battleship-gray F-30 registered a maximum belt horsepower of 32.8 at 1,150 rpm and 24.85 horsepower on the drawbar. Maximum drawbar pull was 4,157 lb. at 2.24 mph. In November 1936, the beginning of the 1937 model year, IH started to paint their tractors the now-familiar red, making them more visible on country roads.

Hard economic times tempered demand for the big F-30. Nevertheless, IH sold some 683 of the tractors in 1931. Production

1935 Farmall F-30 Cane Special
This is the first Cane Special F-30, Serial Number FB7262NW (narrow rear tread, wide front). F-30 production began in 1931 as the first real variation on the Farmall theme. More than 29,000 were built by the end of production in 1939. The F-30, which was considerably larger and more powerful than the Farmall Regular, quickly won the hearts of farmers with larger acreage. Its engine displaced 284 ci (compared to the Regular's 221 ci). Owner, collector, and restorer John Wagner had to use 46-inch rear wheels, as the original, however.

steadily increased over the next decade, peaking at 8,502 in 1937. When production ended in 1939, more than 29,000 had been built. IH made almost no changes to the basic design over the eight-year production run. Harvester offered versions of the F-30 that featured narrow and wide rear tread, wide and narrow front ends, and a Cane, or high-crop, version. Rubber tires and electric lights were options. The Farmall F-30 sold for between $1,100 and $1,300.

COMPARISONS

Make/Model	CID	Cyl.	Belt/PTO HP	Drawbar Pull (lb.)	Fuel Cons.*	Weight (lb.)
Farmall F-30	**283.7**	**4**	**32.8**	**4,157**	**9.61**	**5,990**
Oliver 18/27	280	4	29.72	3,664	9.7	3,825

*horsepower hours per gallon
Weight in pounds is as tested including ballast (if any) and driver.

1936 Farmall F-30
The F-30 was a true three-plow tractor designed to appeal to the farmer with more land to till. The F-30 could also drive a large thresher with almost 33 horsepower on the belt. Farmalls were painted gray until November 1936, when International Harvester introduced the familiar Farmall Red on its 1937 models. *Ralph W. Sanders/ Motorbooks Archive*

1935 Farmall F-30
This F-30 (narrow front, wide rear) is equipped with stylish rear fenders and F&H spoke wheels, both of which were factory options, as well as pinstriping. The fancy exhaust collar was also an option.

1937 Farmall F-30
For the 1937 model year, Farmall began painting their tractors their now-familiar red, making them more visible on country roads. *Ralph W. Sanders/Motorbooks Archive*

The Farmall F-20

Production of the F-20, which replaced the Farmall Regular, began in 1932. The F-20, which weighed 4,500 lb., was slightly longer and heavier than the Regular. Improvements to the F-20's head and pistons coaxed about 10 percent more horsepower out of the same 221-ci engine. Nebraska tests in April 1934 indicated a maximum belt horsepower of 23.11 and a maximum drawbar horsepower of 15.38. It is interesting to note that John Deere's Model A was the next tractor to be tested. The Model A weighed less at 4,059 lb., had slightly more maximum belt horsepower at 24.71, and more drawbar power at 18.72. The F-20's fuel economy under maximum load was slightly better at 10.41 horsepower hours per gallon versus 10.1 horsepower hours per gallon for the two-cylinder Deere.

F-20 options included rubber tires, wide front ends, and adjustable wide front ends. The tractor cost between $895 and $1,000. The F-20 was IH's most popular tractor until production ended in 1939. Almost 150,000 were sold during the F-20's nine-year production run.

COMPARISONS

Make/Model	CID	Cyl.	Belt/PTO HP	Drawbar Pull (lb.)	Fuel Cons.*	Weight (lb.)
Farmall F-20	**221**	**4**	**23.11**	**2,334**	**10.41**	**4,545**
John Deere A	309	2	24.71	2,923	10.1	4,059

*horsepower hours per gallon
Weight in pounds is as tested including ballast (if any) and driver.

1936 Farmall F-20
A Farmall F-20 stands next to a 1928 International 1-Ton "6-Speed Special" (a three-speed transmission and a two-speed axle) truck.

(*opposite page*)
1936 Farmall F-20
The F-20 was an improved version of the Farmall Regular, featuring about 10 percent more power, even though the engine displacement remained the same at 221 ci. Harvester used a new four-speed transmission on the F-20 instead of the three-speed unit on the Regular. The F-20 pictured here has the "narrow" rear axle.

A young farmer from Duluth, Minnesota, learns how to operate the Farmall F-20. *Photo courtesy of Minnesota Historical Society*

1933 Farmall F-20 Duckbill
The F-20 replaced the Regular in 1932. F-20s produced before mid-1934 were equipped with steering gears that were much the same as on the Regular, except they were covered with this odd-shaped housing. Later F-20s featured worm and wheel steering gears, which greatly reduced shock feedback to the driver. Many early F-20s have been converted to the worm gear steering. Original F-20s have become known as the "duckbills" because of the tractor's unique steering gear housing. Some early F-30s also had duckbill steering gear boxes.

1939 Farmall F-20
The last year for the F-20, this version had many of the features of the subsequent Model H, including foot brakes on the right side of the platform that could be locked together. John Wagner's father originally bought the tractor in 1945 but then sold it in 1947 when he purchased a new Farmall MD. John was able to buy it back after a dealer had installed an electric starter and generator. Installing the starter was no small task. The dealer had to fit a ring gear to the flywheel and modify the bell housing.

F-20
McCORMICK-DEERING
FARMALL

F-20
McCORMICK-DEERING
FARMALL

F-20
McCORMICK-DEERING
FARMALL

F-20
McCORMICK-DEERING
FARMALL

1938 Farmall F-20
This wide-front beauty features the optional half rear fenders and an aftermarket starter and generator. A muffler was also an option on this model, although buyers seldom added them. For an extra $130, a farmer could add the adjustable wide-front option. The wide-front version was only available with the narrow rear tread, however. One of the most important patented features of the early Farmalls was the steering wheel–cultivator gang shift. The lever that operates the gang shift protrudes from the front A frame. The F-20's fixed drawbar featured a variety of holes to suit most any requirement. International Harvester began to install POT safety shields as an option toward the end of the 1930s.

Author Robert N. Pripps takes the 1938 F-20 for a test drive.

Farmers in Box Elder City, Utah, pause for a photo in front of their F Series Farmall, circa 1940. *Library of Congress*

The Farmall F-20 made harvest time much more manageable.

The Farmall F-12

In early 1931, one of Alexander Legge's last acts was to steer the New Works Committee away from creating larger and more expensive tractors and convince them to build versions that would appeal to the small farmer who still farmed with horses. He was quoted as saying that the economic conditions of the 1930s were much the same as they had been in the 1920s, when the inexpensive Fordson rocked their empire. Studies resulted in a one-plow tractor of about ten horsepower, known as the F-10.

In May 1932, a prototype emerged for testing that used a semi-unit frame with stub rails for the front end. The designers took a page from the Oliver notebook and built the F-12 without the bull-gear drop box at the ends of the axles, using larger-diameter wheels to get the desired crop clearance instead. (Oliver and others had adopted this arrangement for their GP tractors as early as 1929.) This feature allowed the use of splined axle extensions, which made it easier to adjust the tread width.

Originally, the F-12 was equipped with a Waukesha engine, but IH soon replaced it with one of its own. Their four-cylinder OHV engine displaced 113 ci and was rated at 1,400 rpm. The F-12 was available with a gasoline or distillate fuel system. On distillate, maximum belt horsepower was 14.59 and maximum drawbar horsepower reached 11.81. The little tractor, weighing in at 3,280 lb., had a three-speed transmission rather than the four-speed unit on the F-20 and F-30. Rubber tires and the wide front end were optional.

International Harvester built more than 120,000 of the F-12s between 1932 and 1938. Fairway versions were interspersed in the production run. At $600, the F-12 cost about a third less than an F-20 but still gave the small farmer all the advantages of an all-purpose tractor.

1932 Farmall F-12 (Pre-Production)
F-12 production began in 1932, with only 32 built that year. After that, production took off, with more than 123,000 built by the time it was replaced by the F-14 in 1938. *Ralph W. Sanders/ Motorbooks Archive*

1937 Farmall F-12
Though dual tricycle front wheels were standard on the F-12, both a single front wheel and a wide front axle (seen here) were also available. *Ralph W. Sanders/ Motorbooks Archive*

The Farmall F-14

In early 1938, the F-14 replaced the F-12. The F-14 was essentially the same tractor, but it had a raised steering wheel, which provided a more comfortable angle for the driver. The revised steering wheel position had actually been available aftermarket on the F-12 for some time.

The governor on the F-14 was reset to 1,650 rpm, rather than the 1,400 rpm setting for the F-12. This adjustment increased the belt horsepower from 14.59 to 17. Maximum drawbar power was 13.24 horsepower. Kerosene fuel powered the F-14 in these ratings, but gasoline setups were optional for both the F-14 and the F-12. With the higher compression heads, gasoline provided substantially more power.

Production of the F-14 lasted for only two years. By mid-1939, the beautiful Raymond Loewy–designed letter series Farmalls were on the scene.

COMPARISONS

Make/Model	CID	Cyl.	Belt/PTO HP	Drawbar Pull (lb.)	Fuel Cons.*	Weight (lb.)
Farmall F-12	**113**	**4**	**14.59**	**1,814**	**10.01**	**3,280**
Farmall F-14	**113**	**4**	**17.0**	**2,369**	**10.86**	**4,900**
John Deere H	99.67	2	14.22	1,839	11.95	3,035

*horsepower hours per gallon
Weight in pounds is as tested including ballast (if any) and driver.

The Farmall F-14 made tending the fields much easier while American men were overseas fighting World War II.

1939 Farmall F-14
The F-14 replaced the F-12 in 1938. An increase in engine speed (1,400 to 1,650 rpm) gave the F-14 a two-plow rating. A higher steering wheel provided a more ergonomically correct riding position, and a hydraulic implement lift was an option but almost universally included. *Ralph W. Sanders/Motorbooks Archive*

1939 Farmall F-14

Production of the F-14 began in 1938 when Harvester discontinued the venerable F-12. The two were almost identical, except the F-14's engine governor was set to 1,650 rpm rather than the F-12's 1,400 rpm. The F-14 featured a raised steering wheel, providing the driver with a more comfortable angle. This feature had been available as an upgrade for the F-12 for some time. The increased engine speed gave the F-14 about a 9 percent horsepower advantage over the F-12, enabling it to handle two 14-inch plows. The F-14 was available in orchard, Fairway, tricycle row-crop, and adjustable wide-front row-crop versions. Here, owner Ron Hattendorf drove his wide-front row-crop with Kleber round-spoke rear wheels. This tractor was originally sold by Bunge Hardware, an IH dealer in northern Illinois. Hattendorf preserved the dealer sticker that was originally affixed to the tractor.

SPECIFICATIONS

Model	Regular	F-30	F-20	F-12 (G)	F-12 (K)	F-14
Years	1924-1932	1931-1939	1932-1939	1932-1938	1932-1938	1938-1939
Test No.	117	198	221	212	220	297
Test Date	Sep. 1925	Oct. 1931	Apr. 1934	May 1933	Nov. 1933	Apr. 1938
Engine Cyl./CID	4/221	4/283.7	4/221	4/113	4/113	4/113
Rated RPM	1200	1150	1200	1400	1400	1650
Drawbar HP	12.7	24.85	15.38	11.81	11.81	13.24
Belt/PTO HP	20.05	32.8	23.11	14.59	14.59	17
Drawbar Pull (lb.)	2,725	4,157	2,334	1,870	1,814	2,369
Test Weight (lb.)	4,100	5,990	4,545	3,280	3,280	4,900
Fuel	Kerosene	Kerosene	Kerosene	Gasoline	Kerosene	Distillate
Transmission	3f-1r	4f-1r	4f-1r	3f-1r	3f-1r	3f-1r
Test Configuration	RC-Tri	RC-Tri	RC-Tri	RC-Tri	RC-Tri	RC-Tri
Top Road Speed (mph)	4	4	4	4	4	5
Price, New ($)	825-950	1,100-1,300	895-1,000	600-700	600-700	800-850
Production	31,000	28,500	130,000	—— 120,000 ——		27,000

1949 Farmall Cub.
Ralph W. Sanders/Motorbooks Archive

CHAPTER 3

Farmalls with Style 1939–1953

"True strength is delicate."
—Louise Nevelson, artist, 1939

The continuing effects of the Great Depression forced durable-good manufacturers to find creative ways to spark the sales of their products. Beginning in 1933, some companies were able to boost sales through styling—a method of making a product's design more appealing and its form more functional, while clearly differentiating their product from a competitor's. Such changes reflected a certain optimism about future economic conditions. The automobile industry was among the first to employ the styling technique. Until the early 1930s, cars were very similar to one another: They were boxy with four-cylinder engines, exposed radiators, and clamshell fenders. By the mid-1930s, fender skirts; slanted and divided windshields; fancy radiator grilles; and six-, eight-, twelve-, and even sixteen-cylinder engines began to appear on new cars. Styling made its way into other products, from telephones to trucks.

Tractor manufacturers approached styling with caution. The Peoria Tractor Company, for example, tentatively introduced the Streamline in the 1920s. The crusty and ever practical farmer, however, tended to resist frills if he thought it unnecessarily increased the price, or made the tractor harder to service. But when Oliver Hart-Parr introduced the Oliver 70 tractor in 1935, it took farm implements to a whole new level. The stylish 70 was so carlike that it immediately overshadowed all competition. Aside from its classic grille and louvered side panels, the Oliver 70

1950s Farmall Cub advertisement.

sported a radiator cap in the form of a shiny, miniature, middle-buster plow. But the tractor was more than just a pretty face—it offered a smooth-running high-compression (for 70 octane gasoline—hence the name), six-cylinder engine, a four-speed gearbox for row-crop versions, six speeds for standard treads, a starter, lights, and a muffler. Advertisements of the day showed "Sister" and "bud" taking their turns at the wheel. Most importantly, because the Oliver 70 was a hit with the farmer, it forced the other manufacturers to respond.

In 1935, engineers at IH were working on their next generation of tractors, focusing mainly on adapting the straight rear axle featured on the F-12 and F-14 to create a version of the F-20. Because of the success of the Oliver 70, International Harvester brought in famous industrial designer Raymond Loewy as a styling consultant. (Loewy later developed the rakish but attractive 1953 Studebaker cars.) The new tractor models incorporated various Loewy stylized features, and gradually began to resemble the final configuration. One version included louvered hood side panels popular at the time. Loewy later vetoed the side panels, arguing that they were nonfunctional and impeded access to the engine.

Meanwhile, competition began to heat up. In 1937, as some of the other tractor makers gained parity with the original 70, Oliver restyled the Model 70 to an even sleeker design. In 1938, the very stylish Minneapolis-Moline UDLX and the Graham-Bradley 503 appeared on the scene. While the last of the F Series Farmalls continued to sell fairly well, the F-20 and F-30 were now up against John Deere Models A and B, newly styled by Henry Dreyfuss. The F-14 also had some competition—the Allis-Chalmers Model B, a nifty little outfit that performed well, looked good, and sold for a mere $600.

But Raymond Loewy was up for the challenge. International had commissioned him to overhaul the entire Harvester product line, from the company logo to operator ergonomics. In his 1979 book, *Industrial Design*, Loewy stated that he thought the original logo "frail and amateurish" for a company with the prestige and power of International Harvester. The new logo, a capital "H" with a dotted "I" in its center, suggests the front view of a row-crop tractor, complete with operator, when viewed head-on.

In 1938, IH introduced the new TD-18 crawler tractor, the first of the models designed by Loewy. With bright red, smoothly contoured sheet metal and a purposeful grille, the TD-18 had undeniably striking good looks. Best of all, the TD-18 was functional, too. Within a few months, the IH line of styled wheeled tractors followed the TD-18, carrying the styling motif further. There were three sizes at first—the A, which replaced the

F-14; the H, replacing the F-20; and the M, replacing the F-30. The 1939 Loewy Farmalls were an instant success and are still quite smart looking today.

Aside from making the International Harvester line more attractive, Loewy made the tractors easier to use. He aligned the instruments so the farmer could check them quickly and without distraction, placed levers and pedals where they were easy to reach and operate, and added new safety features (fenders became standard).

In June 1939, just after the Loewy Farmalls were introduced, Henry Ford sent another shock wave through Harvester when he unveiled the Ford-Ferguson. The new tractor, built in cooperation with Harry Ferguson of Ireland, was a squat combination of a standard-tread and a row-crop. At first, tractor gurus dismissed the challenge as merely a styled Fordson—even the name "Ford-Ferguson" suggested "Fordson." Many even referred to the little gray tractors as "Fordsons," and continue to call them that to this day. When Ford and Ferguson demonstrated the new tractor for the press, however, experienced tractor men could not believe what they were seeing. The small, 2,500-lb. tractor could plow an acre per hour—with an eight-year-old boy at the wheel! It was Ferguson's three-point implement hitch, complete with a hydraulic lift system that automatically compensated for increasing draft loads by slightly raising the implement, that made this unbelievable performance possible. The weight of the implement plus draft and suction loads all increased the load on the rear tires, greatly adding to the traction.

TAKE LIFE EASIER WITH FARMALL-A

The one-row Farmall-A shown with adjustable front axle, available at slight extra cost.

THE Farmall-A is a 1-row tractor that will plow 5 to 7 acres a day and harrow, plant, cultivate, or do any other power job on the small row-crop farm. It has the power to plow deep, full-sized furrows. It has the right wheel treads and the right overall dimensions for work in small or irregularly shaped fields and for planting and cultivating 1 to 6 rows of various spacings.

Its snappy ten-mile high speed adapts it to farm and highway hauling, and it's as easy to drive as the family car.

The One Row Tractor with CULTI-VISION...

THE operator of a Farmall-A and any of the cultivators built for direct attachment to the tractor has the work directly in front of him. He can set his inside shovels as close to the rows as it is safe to go without danger of cutting the plant roots. He can cultivate at speeds up to nearly five miles an hour—14 to 17 acres a day—twice as much as he could do with horses. CULTI-VISION brings clean, thorough cultivation with driving comfort you will find in no other one-row tractor.

Farmall-A is a quality 4-cylinder tractor built to supply all the power needed on the small farm using 2 [illegible] horses. On the larger farm the whole sea[illegible] through Farmall-A will work right along with the larger units, providing unbeatable economy in such operations as planting, cultivating, and haying. Farmall-A is ready to go as regularly equipped. Starting, lighting, and other special equipment is available to meet individual needs.

Farmall-A with lighting equipment, swinging drawbar, belt pulley, power take-off and muffler.

4

A page from an early '40s Farmall catalog suggests that buyers "Take life easier with Farmall-A."

The Farmall A

The Model A was the first and the most radical of the Loewy-styled Farmalls to be released. The Model A used the same engine as the F-14, but it was governed at 1,400 rpm, rather than 1,650 rpm. The Farmall A's size and power placed it in the same class as the new Ford-Ferguson. The A was heavier and had better traction when the Ford was not benefiting from the weight transfer effects of the Ferguson lift system. With this traction-boosting feature, however, the Ford obtained a two-plow rating, while the Farmall A was only rated for one 14-inch plow.

The A had a unique configuration—the engine was offset to left while the operator sat on the right, giving the driver an unobstructed view of the ground beneath the tractor to where the cultivator operated. This feature, called "Culti-Vision," made the Farmall A ideal for cultivating delicate truck garden crops. Although the Ford-Ferguson's rear-mounted cultivator was ahead of its time, farmers rejected the idea when it was first introduced. It took years before farmers would accept cultivators positioned behind the tractor, where the driver could not see them as they worked. Thus the Farmall A, which sold for about $575 when introduced, sold well in its intended market.

All Farmall As featured wide front axles with kingpins that extended downward. The rear axles had geared drop boxes, and the combination of the two provided adequate crop clearance. Standard equipment also included a muffler, starter, fenders, a PTO, and a belt pulley, which was located in the back, rather than on the side per the usual configuration. Farmers chose between gasoline or distillate engines, most favoring gasoline. The gasoline version of the 113-ci four-cylinder OHV engine used a 6:1 compression ratio, producing more horsepower than the distillate version, which carried only 5:1. The Model A originally used a gravity-type cooling system (called thermo-syphon), but Harvester added a water pump on later models. The A featured two-wheel independent brakes and a four-speed transmission. The build of the Farmall AV (high-crop) was similar to the A, but it had longer kingpins and larger rear tires, giving it a crop clearance of six more inches.

(opposite)
1941 Farmall A
Farm tractor styling became paramount in the late 1930s. International Harvester hired famous industrial designer Raymond Loewy (later famous for his sporty Studebaker car designs) as a consultant. The Farmall A was the first of Loewy's wheeled tractors to appear in 1939, and the most radical with its offset Culti-Vision design. Harvester manufactured gasoline and kerosene versions of the A. Standard equipment involved a muffler, a starter, rubber tires, and a four-speed transmission. ***Ralph W. Sanders/Motorbooks Archive***

COMPARISONS

Make/Model	CID	Cyl.	Belt/PTO HP	Drawbar Pull (lb.)	Fuel Cons.*	Weight (lb.)
Farmall A	**113**	**4**	**16.86**	**2,387**	**11.97**	**3,570**
Ford 9N	119.7	4	23.01	2,236	9.74	3,375
A-C Model C	125	4	21.83	2,352	11.86	3,205
John Deere H	99.7	2	14.22	1,839	11.95	3,035
Case VC	124	4	23.76	2,798	11.33	4,290

*horsepower hours per gallon
Weight in pounds is as tested including ballast (if any) and driver.

McCORMICK-DEERING
FARMALL
A

1950 German-Built Farmall FG
The FG, built in the German Neuss factory, was a modernized version of the Farmall F-12, manufactured with stylish sheet metal. The FG was roughly equivalent to the American Farmall A. The four-cylinder engine displaced 125 ci (2,043 cc)—somewhat more than the F-12's 113 ci (1,851 cc), but similar to the Super A-1's 123 ci (2,015 cc). On distillate fuel, the FG had a 20-horsepower rating. Electric start (on gasoline) was standard. In this photo, owner Johann Hood drove the FG while Patrick Kastner rode the Neuss-built McCormick-Deering hay rake.

1960 French Farmall 135-D
Production of the 135-D started in 1957 at Saint-Dizier in France. With a two-cylinder diesel rated at 17 horsepower, the 135-D is comparable to the Farmall A and the Cub. The diesel displaces 1,217 cc (74 ci). The 135-D used many German parts, including a six-speed ZF gearbox.

1958 German-Built Farmall D320
With a 1,631-cc (100-ci) diesel, this German equivalent to the Farmall A is rated at 20 horsepower.

1952 German-Built Farmall DF-25

Roughly equivalent to the U.S. Farmall H, the DF-25 had a four-cylinder diesel engine of 2,043 cc (125 ci), rated at 25 horsepower. The DF-25 was built in the IH factory in Neuss on the Rhine River. The factory was destroyed during World War II but was rebuilt and was manufacturing tractors again by 1947. The DF-25 was built from 1951 to 1953. The engine is similar to the FG gasoline/distillate engine, but heavily strengthened. Patrick Kastner drives the DF-25 with Johann Hood on the McCormick 8B mower (also built at Neuss).

The Farmall H

In 1939, about a month after the A became available, Harvester introduced the Model H Farmall. The Model H replaced the venerable F-20, which had taken the place of the original Farmall, now called the "Regular." International Harvester built seven examples of the Model H in 1938, referred to at the time as the Model 2-F. The Farmall H, which became the most popular Farmall, was rated for two 14-inch plows. International Harvester also produced a cane version, called the HV, which provided a crop clearance of 30.5 inches. The H had a spacious platform and a comfortable seat. Both wheel brake pedals were on the right side and could be hooked together as a service brake. By the time the Model H came to be, Harvester no longer installed the cable steering brakes of the F-Series on its tractors.

International Harvester took a humorous approach in this ad for the Farmall H, B, and A.

The Farmall H is the center of attention at the fair.

Harvester engineers developed a modern, four-cylinder overhead-valve engine of 152 ci for the new H. Operating at a rated speed of 1,650 rpm, the new engine placed the H in the 25-horsepower class. The tractor was equipped with a five-speed gearbox instead of the four-speed unit of the F-20. The five-speed transmission remained part of the H throughout the life of the series.

The Farmall H was an overnight success. Its strikingly bold styling still looks modern today. But in 1939 and into the 1940s, the H faced stiff competition from other tractor companies, particularly John Deere with its Model B.

COMPARISONS

Make/Model	CID	Cyl.	Belt/PTO HP	Drawbar Pull (lb.)	Fuel Cons.*	Weight (lb.)
Farmall H	**152**	**4**	**24.28**	**3,603**	**11.7**	**5,500**
John Deere B	175	2	18.31	2,690	10.56	5,000
M-M RTU	165	4	21.63	2,666	10.14	3,400
Case SC	165	4	21.62	3,166	10.55	4,200
Massey 101 Jr.	140	4	30.15	3,278	11.75	4,825

*horsepower hours per gallon
Weight in pounds is as tested including ballast (if any) and driver.

1948 Farmall HV
Jeremy Sevcik polishes his father Norm Sevcik's high-crop Farmall H. This tractor, with the wishbone wide front end, came from the cotton country. The front tires are on 20-inch rims, making the tires hard to find and expensive.

1940 Farmall H
Harvester introduced the H right after the Farmall A in 1939. Raymond Loewy designed the aesthetically pleasing and functional tractor to compete with the John Deere Model B, which was unveiled in 1938 with styling by Henry Dreyfuss. IH engineers developed a new engine for the H, a modern high-speed unit of 152 ci operating at 1,650 rpm. It used water-pump cooling, rather than the thermo-syphon system of the F-20. The H was available in both distillate (kerosene) and gasoline versions. *Ralph W. Sanders/Motorbooks Archive*

1942 Farmall H
The Farmall H and M used the same wheelbase so that cultivators and other body-mounted implements could be interchangeable. Many wartime Farmalls were equipped with steel wheels. *Ralph W. Sanders/ Motorbooks Archive*

1940 Farmall H

This Model H, serial number FBH22130, was built with steel wheels, a rarity for that time period. The H replaced the F-20 in the Harvester lineup and soon became the most popular model. Built without a starter, generator, or implement lift, this Farmall H sold for about $750 in 1940. One with full rubber, electrical system, and lift would have been about $200 more. The four-cylinder OHV engine produced about 25 horsepower on gasoline (a distillate version was also available with slightly less power due to a lower compression ratio), enough for two 14-inch plows, or a 22-inch thresher.

The Farmall M

With the unveiling of the Model H, Harvester also introduced the Model M. The big M was the envy of farmers with lesser tractors. At $1,200, the M cost $200 more than the H, and in the early 1940s, that was a big difference.

The M, which replaced the F-30 in the International Harvester stable, had a lot in common with the Model H. Harvester built both tractors on the same type of frame, equipped with holes to mount cultivators. Both the Model H and the Model M were available in high-crop versions, with drop gearboxes at the ends of the rear axles. These two tractors were available with adjustable wide front ends, although farmers tended to prefer dual tricycle fronts. On both models, buyers had the choice of a kerosene or

1940 Farmall M
The M, which replaced the F-30, had a lot in common with the Model H, but farmers who chose the M were paying for a bigger engine with two-thirds more power than the H. This example is hooked up to a No-Till planter. *Ralph W. Sanders/Motorbooks Archive*

— FOWLER McCORMICK —

Fowler McCormick, born in 1898, was the grandson of both Cyrus Hall McCormick and John D. Rockefeller, the oil magnate. His father was Harold McCormick and his mother was Edith Rockefeller McCormick. He was named for his grandmother, Nancy Maria (Nettie) Fowler McCormick.

By the time Fowler was born, the McCormick family was considered "the old rich." And, of course, the Rockefellers were old rich. In such comfortable surroundings, young Fowler did not take work or schoolwork too seriously. When the death of an older brother left his mother debilitated with depression, Fowler dropped out of Groton School in Massachusetts and went with her to Switzerland. There, his mother spent many hours with Dr. Karl Jung, the famous psychiatrist, while Fowler attended school. When he returned to the United States, Fowler McCormick went to Princeton. Here, living the life of a rich young gentleman, he fell in love with Anne Urquhart Stillman, the mother of one of his friends and a woman nineteen years his senior. Anne went by the nickname Fifi. She eventually divorced her husband to marry Fowler.

Fowler was still not ready to settle down to a career but dabbled in music, finance, and psychology. In 1926 at the age of twenty-eight, however, Fowler began training incognito at the McCormick Milwaukee foundry. The six months he spent loading pig iron into a furnace had a settling effect on him. When Alexander Legge invited Fowler to enter the management training program, he jumped at the chance. As he rose through the ranks, Fowler became known for fostering a McCormick family feeling among his associates and employees. This went so far as to be anti-union. As Fowler neared the top of the corporate ladder in the 1930s, management instituted employee work councils as a substitute for labor union recognition. This led to an extended period of labor unrest and crippling strikes up until World War II. Fowler McCormick succeeded his father, Harold McCormick, as IHC President in 1941. (Harold had taken the company reins in 1918 from his brother, C. H. McCormick II. C. H. II, or "Junior," as he was sometimes called, took over from the inventor of the reaper in 1881.)

After the war, national unions had a firm foothold in Harvester, but the friction between labor and management was always just below the surface. Excellent postwar business put off the inevitable until the labor strife of the 1980s forced the merger with Case. Yet it was said that the ultimately fatal labor strife began with Fowler McCormick's paternalistic policies.

Fowler's close associate and longtime friend John McCaffrey succeeded him as board chairman. He was the first non-family member in the top job. Unfortunately, McCormick and McCaffrey soon began to disagree on company policy. Eventually, McCormick delivered an ultimatum to the board regarding conflicting policies between him and McCaffrey. The board sided with McCaffrey, and in 1951, Fowler was out. Fowler McCormick blamed John McCaffrey for the company's loss of standing during his administration. These allegations were at least partly true, as McCaffrey, who had become president when Fowler became board chairman, contributed to much of the labor unrest. McCaffrey was also the one who expanded the number of models and types of trucks to an unmanageable level.

gasoline-fueled engine (the gasoline version was the most popular on both tractors). Except for a period during World War II when rubber was in short supply, Models H and M were usually equipped with rubber tires. The H and M both featured a rear-axle design, pioneered directly from the differential housing. A keyed axle shaft allowed the operator to make infinite adjustments to the rear-wheel tread width. With the exception of high-clearance versions of these tractors, neither model had drop gearboxes.

Farmers who chose the Model M were paying for a bigger engine with more power; the M had two-thirds more horsepower than the H. The M's new 248-ci overhead-valve four-cylinder engine with magneto ignition performed at a normal operating speed of 1,450 rpm. The M was available with either high-compression heads or low-compression heads with a kerosene manifold. The M featured a five-speed transmission with an optional "Life-All" hydraulic system. Rated for three 14-inch bottoms, the M was available with a wide front end or with the dual tricycle setup. International Harvester also produced the high-crop versions of the Model M with wide or single-wheel fronts.

1945 Farmall M
Like a monument in an unattended graveyard, this derelict Farmall M attests to a long life of hard work.

In 1941, Harvester surprised the industry when it introduced the MD, a diesel engine version of the M. International Harvester had pioneered diesel wheeled tractors when they introduced the WD-40 in 1934. Caterpillar had unveiled the first diesel tractor, the Caterpillar Diesel Sixty, in 1932. By 1941, diesel engines were used in stationary power plants, ships, crawlers, and even dirigibles, but they were not yet common to trucks or other wheeled vehicles. Caterpillar introduced its DW-10 diesel wheeled tractor, teamed with an earth mover scraper in 1940, but the Farmall MD was the first diesel-powered row-crop farm tractor.

The diesel's 248-ci four-cylinder OHV engine block was the same as the one used in the gasoline-powered M. But the diesel version featured aluminum pistons and five main bearing cranks instead of three. The main difference between the diesel and the gasoline engine was in the head. Besides having twice the compression ratio of the gasoline version, the diesel engine head contained the IH starting system. Gasoline was used to start the engine, and when the engine warmed up, the operator could switch the engine over to diesel power using a lever on the platform. Pulling the lever for starting on gasoline triggered a chain of events. First it enlarged the combustion chamber to provide a compression ratio of 6.4:1, then intake air came in through the carburetor and gasoline flowed into the carburetor, completing an ignition circuit, which provided fire to the spark plugs. After a minute or so on gasoline, actuation of the lever shut all that off and compression increased to 14.2:1. With the diesel "throttle" open, the engine then ran as a diesel.

Large-acreage farmers found the MD, which used about a third less fuel at full power, paid for its initial cost over a conventional M in less than a year. And in those days, diesel cost a little over half as much as gasoline. Fuel consumption savings were even greater over gasoline and distillate at part loads.

A painting of a Farmall M from a 1971 International Harvester calendar.

The Farmall M was the first tractor to be assembled in Harvester's plant in Doncaster, Yorkshire, England. International Harvester was indeed "International." Various plants had been established in Europe and South America. Tractors had traditionally been exported from the United States until assembly of the M in Doncaster. Production of the BM, or British Model M, began in 1949, some ten years after the M was introduced in the United States. The Farmall BMD diesel came out in 1952. The Doncaster plant also made a wide variety of IH implements.

COMPARISONS

Make/Model	CID	Cyl.	Belt/PTO HP	Drawbar Pull (lb.)	Fuel Cons.*	Weight (lb.)
Farmall M (gas)	**247.7**	**4**	**36.07**	**4,233**	**12.16**	**6,770**
Farmall M (dist.)	**247.7**	**4**	**34.16**	**4,365**	**12.50**	**6,770**
Farmall MD	**247.7**	**4**	**35.02**	**4,541**	**14.57**	**7,570**
Case DC	260	4	35.51	4,128	12.13	7,010
John Deere A	309	2	28.93	4,110	11.3	6,410

*horsepower hours per gallon
Weight in pounds is as tested including ballast (if any) and driver.

1941 Farmall MD
The diesel-powered Farmall M came out in 1941. Although it cost one and a half times as much as the gasoline version, fuel consumption was about 33 percent less. This model has aftermarket power steering, live hydraulics, live PTO, and an auxiliary gearbox. Owner Alan Smith, the famous Caterpillar collector from McHenry, Illinois, was at the controls for this photograph.

1949 British-Built Farmall M
This M, serial number 1027, was the seventy-second one built at the Doncaster plant. Later that year, Doncaster Ms were labeled "BM," meaning "British made."

1949 Farmall M
Myles Sevcik bought this Farmall M new with the Model 31 McCormick loader. The M, introduced in 1939, was the broad-shouldered version of the classic H. Harvester engineers designed a brand-new four-cylinder OHV engine for the M of 248 ci, which operated on either gas or distillate (there were different compression rations for each fuel). The 1949 Farmall M was rated for three 14-inch plow bottoms.

1948 Farmall MV
This single-front-wheel high-clearance version of the famous Farmall M came from the cotton fields of the South. It now, however, resides in Minnesota at "High-Crop Hill" near Northfield, its cotton picking days over.

In this ad featuring a Farmall M, International Harvester advises customers to have their Harvester tractors serviced before the spring rush.

Some thoughts on Food Production...and Safety

International Harvester has received permission to use the statements quoted on this page, and extends its thanks to:
GMC TRUCK & COACH DIVISION of General Motors Corporation
THE PENNSYLVANIA RAILROAD
and NEWSWEEK MAGAZINE

1 "In Napoleon's time, it took nineteen farm workers to supply food for one person off the farm. Today, one farm worker provides food for four hungry people elsewhere.

"Early in the last century, it required sixty-four hours of work to produce an acre of wheat. Now it can be done in less than two and one-half hours.

"This means that the record crops produced in World War II were grown and harvested by 26 million fewer persons than would have been required if early 1800 methods were still in use."

—GMC TRUCK & COACH DIVISION of General Motors Corporation

• • •

2 "When the Pennsylvania Railroad was founded in 1846, it took three-quarters of the total population to produce food and fiber to feed and clothe themselves and the other one-quarter living in towns and cities.

"Today 25 per cent do that job—and do it better!"

—THE PENNSYLVANIA RAILROAD

THE items at the left, Nos. 1 and 2, from statements published recently, bring a feeling of pride to International Harvester. They remind us of this Company's part in the great record of American agriculture.

The seventy thousand men and women who devote their lives to service under the IH symbol think back to 1831, to Cyrus Hall McCormick and the First Reaper. They think of what the engineers and builders have accomplished in the 115 years that have followed. They are proud of the part this company has played in the advance of power-and-machine-farming—and of what it is building today.

Item No. 3, at the right, is different. It carries a message of warning for every farmer...In the coming year International Harvester will do its utmost to provide all farmers with the machines they need. These machines carry many safeguards for safety. Nevertheless, when you use this equipment: Take Time to be Careful. Yours for better living on the family farm.

INTERNATIONAL HARVESTER COMPANY
180 North Michigan Avenue, Chicago 1, Illinois

3 Safety on the Farm: "'Farming is today one of the most hazardous of all occupations,' warned Dr. H. H. Young and Dr. Ralph K. Ghormley of the Mayo Clinic. Making a nine-year study of farm accidents treated at the clinic, they found these included at least 65 serious cases each year. (Note: Of the 17,500 occupational accident deaths in the United States in 1944, 4,300, or 25 per cent, involved farmers.)

"In this series, falls led all other causes of accidents—most commonly falls from some piece of farm equipment—and they caused a mortality rate of 5 per cent. Second place, with 186 victims, were accidents from farm machinery. The third, with 104 cases, were accidents caused by livestock. Although the bull is generally supposed to be most dangerous, horses accounted for most of the injuries—66 accidents.

"'The first step in prevention is education in safety methods,' the Mayo doctors said. 'The operation of farm equipment demands as much understanding and respect as the running of a moving locomotive.'"

—NEWSWEEK MAGAZINE, July 15, 1946

HEAR JAMES MELTON • "HARVEST OF STARS" EVERY SUNDAY • NBC NETWORK

GREAT THINGS ARE COMING IN THE FARMALL SYSTEM — INTERNATIONAL HARVESTER

INTERNATIONAL HARVESTER

IH stresses safety in this ad featuring the Farmall M.

Farm Security Administration (FSA) officials examine a Farmall M at work near Waterloo, Nebraska, in 1940. *Marion Post Walcott/ Library of Congress*

1949 Farmall M LPG
A liquefied petroleum gas (LPG) fuel option was not offered by IH in 1949 but was a popular aftermarket conversion in the 1950s. LPG burned very clean, so it extended engine life dramatically. Although LPG had a lower heat value (BTU/lb.) than gasoline, the increased compression ratio of the converted engine more than made up for it.

1941 Farmall MD
The Farmall M replaced the F-30 in 1939. The diesel version became available the following year, featuring a complex starting system that switched the combustion chambers from diesel to gasoline for starting. When the engine was warmed up, it switched back to diesel. *Ralph W. Sanders/ Motorbooks Archive*

1949 Farmall MD
Owner John Wagner drives his Farmall MD, a wide-front-end version that was one of the 12,000 Ms made at International Harvester's Louisville plant. The MD used the same engine block as other Ms. It did incorporate additional crankshaft bearings to take the increased BMEP (brake mean effective pressure).

1948 Farmall M Industrial
This yellow, but essentially standard, Farmall M wide-front was used to maintain the highways of Marion County, Kansas. John Wagner of White Pigeon, Michigan, painstakingly restored the tractor.

The Farmall B

The Model B, which rolled off the assembly line in September 1939, was the last of the original prewar-styled Farmall series. Although the Model B looked nothing like the Model A, they were essentially the same tractor. Whereas the offset A featured a wide front with the engine and drive train positioned on the left and the seat and steering wheel on the right, on the B the engine was in the middle, with either a dual tricycle, or single-wheel front end. Unlike the A, which had one long and one short axle, the B had two long axles. The operator's position on the B was like that of the A, offset to the right. The Model As and Bs were built on the same assembly line, and they shared the same serial number sequence.

The B was equipped with either a gasoline or distillate engine, and it was capable of pulling one 16-inch bottom. The basic Model B weighed about 1,900 lb., but its weight could be doubled with ballast. The B was not made in high-crop or industrial versions, and, unlike the A, there were no follow-on versions of the B. Production ended in 1947.

1942 Farmall B(N)
The last of the Loewy-styled Farmalls to come out in 1939, the Model B was closely related to the Farmall A. Whereas the A was "offset" with one long rear axle and one short one, the B had its engine in the center with two long rear axles. The Farmall B was available with either a single front wheel or the dual tricycle, as shown here. The A was only available in the wide-front configuration. The "N" designation signifies a narrower rear tread than the standard B. Instead of a flat belt pulley, a large, V-belt pulley drives the alternator on this Farmall B(N), which carries a 4,000-watt Generac alternator behind its rear axle.

Farmall B(N)

1943 Farmall B ad celebrates Farmall's twentieth birthday.

McCormick's No. 45 Baler ad features the hardworking Farmall Super H and Super C.

The Farmall C

Throughout World War II, IH had struggled with the idea of developing a midsize Farmall. Plans originally centered around a tractor, referred to then as the Model E, with a frame like the H and M, a five-speed transmission, and a live, or continuous, power takeoff (PTO). But, just as in 1924, when IH salesmen feared the new Farmall would steal sales from the McCormick-Deering Model 10-20, faint hearts within the company worried that the Model E would take sales away from the H, which was selling well. Thus, IH dropped plans for the E in favor of the C.

The Farmall C more or less replaced the Model B in 1948. Although it appeared to be much bigger than the B, both featured the 113-ci engine. On the C, however, the engine was rated at 1,650 rpm, rather than at 1,400 rpm as on the A and B models. The difference in appearance came from the use of straight axles and 9x36 rear tires (the A and B used 8x24 rears). All three models employed the same four-speed transmission, but on the C, the differential ratio was changed to accommodate the larger wheels and higher engine speeds.

The C also looked larger because it featured a proper operator's platform, which raised the driver to a higher position. On the C, the seat was centered and the steering wheel was cocked up at an extreme angle compared to the other Farmalls. The steering shaft, with two universal joints, passed along the left side of the engine.

COMPARISONS

Make/Model	CID	Cyl.	Belt/PTO HP	Drawbar Pull (lb.)	Fuel Cons.*	Weight (lb.)
Farmall C	**113**	**4**	**19.91**	**2,902**	**11.24**	**4,409**
Ferguson TE-20	120	4	24.02	2,650	10.57	4,268
John Deere M	101	2	19.49	2,329	11.12	3,850
Massey-Harris 20	124	4	26.08	2,898	11.39	4,490

*horsepower hours per gallon
Weight in pounds is as tested including ballast (if any) and driver.

1949 Farmall C

Almost 80,000 Farmall Model Cs were delivered between 1948 and 1951. The C replaced the Model B in the IH lineup. The two tractors were quite similar, although the C looks much larger due to its larger rear tires. Both the C and the B had the same 113-ci engine and four-speed transmission. An operator's platform on the C puts the driver above the tractor, rather than astride as on the B.

1950 Farmall C (White)
In the early 1950s, IH launched a dramatic marketing program for its small tractor line, painting Cs, Cubs, and Super As a striking white. Collector Dick Alberts of Kent City, Michigan, found this C at an auction. With just a little scrape of his jackknife, Alberts saw white paint under the red. After buying the tractor, he sandblasted it to reveal all white paint underneath. As part of his restoration process, Alberts then had replicas of the original promotional signs copied from old photographs.

The Farmall Cub

The Farmall Cub, which was first introduced to the public in 1947, would always have the distinction of being the smallest Farmall. Plans for the Cub, or "Baby Farmall," began as early as 1943, after IH product planners discovered in a review of 1940 census figures that there were more American farms under forty acres than over forty acres. In designing their Model A and B tractors to target the forty-to-seventy-acre farmer, IH had ignored a vast, untapped market. In analyzing census data, IH decided those most likely to buy the smaller tractor were small tobacco or vegetable farmers with about twenty acres in the eastern and southern states.

Originally known as the Farmall X, a designation most likely used to indicate an experimental model, the new tractor was a smaller version of the successful Model A. In the planning stages, the target price of the new model was $400. IH designers created

1948 Farmall Cub
The Farmall Cub was designed for the under-forty-acre farmer. A four-cylinder L-head engine of 10 horsepower (60 ci) powered the Cub through a three-speed transmission. The Cub's configuration was similar to that of the Farmall A, which was aimed at farmers with forty to seventy-five acres. When designing the Cub, IH engineers were shooting for a target weight of 1,000 pounds (the final product weighed in at 1,500 pounds) at a $400 price point. When Norm Sevcik of Northfield, Minnesota, first bought this 1948 Cub, it smoked a little, so he thought he would have to re-sleeve the little engine. He discovered to his surprise that Cub engines didn't use sleeves. The overhaul was an expensive proposition. Rather than simply replace the cylinder sleeves, Sevcik had to rebore the block and install oversized pistons.

1950 Farmall Cub (White)
The diminutive Cub in white livery startled viewers on showroom floors across America in 1950. It was designed to do just that—to attract attention to the entire Farmall lineup. IH also delivered Farmall As and Cs in white to dealers. Some of these tractors were eventually repainted and sold, while others were sold in white. The Cub's engine allowed a one 12-inch plow rating.

all new parts, which were both cheaper and lighter than those on the Model A. Shooting for a target weight of 1,000 lb., IH considered using a two-cylinder engine for a time but finally overruled the idea because it would look as if IH were copying John Deere. Eventually, the design team settled on a side-valve four-cylinder engine of a shade less than 60 ci.

The IH naming committee settled on the name "Cub" for the smallest Farmall. "Cub" had a friendly ring to it that connoted small, cute, and tough. And airplane manufacturers Piper and Taylor had great success with their small utilitarian airplanes under the same name.

Assembly began in 1947 in the Louisville plant. Target production was set at 50,000 units per year, although this number was never quite reached. The Cub introduced to the public weighed in at about 1,500 lb. without additional ballast and 2,700 lb. as tested at the University of Nebraska. Its initial selling price was closer to $600 than the targeted $400. This was not as much of a problem for IH as originally anticipated, since the price of the new Ford 8N had risen to more than $1,000 following the split with Harry Ferguson. Ford had actually lost money on all of the 9N and 2N models that were sold through Ferguson. Now that "Young Henry" had taken over, altruistic ideas of helping the farmer went out the window.

The Cub's configuration was virtually identical to the Farmall A at 80 percent its size. Unless the two models are placed side by side, it's difficult to tell them apart. The shape of the fuel tank does vary between the two—the Cub's is rounded, while the A's is shaped more like a teardrop. The Cub featured a three-speed transmission.

International Harvester stopped producing the Farmall Cub in 1958, but some Cub Lo-boys were Internationals. IH also produced a regular Cub under the International name from 1958 until 1975.

COMPARISONS

Make/Model	CID	Cyl.	Belt/PTO HP	Drawbar Pull (lb.)	Fuel Cons.*	Weight (lb.)
Farmall Cub	**59.5**	**4**	**9.23**	**1,596**	**10.94**	**2,702**
Allis-Chalmers G	62	4	10.07	1,143	9.06	1,719
Massey Pony	62	4	10.71	1,432	10.33	2,390

*horsepower hours per gallon
Weight in pounds is as tested including ballast (if any) and driver.

1958 International Cub Lo-Boy
IH built Farmall Cubs through 1958 in the regular and Lo-Boy configurations. Harvester released the International Cub in 1958 and manufactured it until 1975 in both the regular and Lo-Boy versions. The mechanics of these models remained much the same through the years, although in 1950, the engine speed was increased from 1,600 rpm to 1,800 rpm, giving the Cub about one more horsepower. The overall height of the Lo-Boy was only 3 inches shorter than the regular Cub, but it had a much lower center of gravity, making it an ideal hillside mower.

1958 French Farmall Cub
Manufactured at the Cima factory in Paris, the French Cub has a four-cylinder engine of 974 cc (60 ci). The French model uses a Solex carburetor, giving it 10.5 horsepower at 1,600 rpm.

1952 Farmall Cub
You know you have a small tractor when you have to chain it down to keep it from being stolen. This one was for sale at a Phoenix gas station. *Photo by R. Pripps*

The Super Series

With the Farmall Super Series tractors, IH added hydraulics and, in most cases, power to their tractors. This line did not include a Super B or a Super Cub. The Super M offered the Torque Amplifier auxiliary transmission, which was a substantial new development.

The Super A

In 1947, IH introduced the Super A to replace the regular A. It was similar to the A, except it featured the new hydraulic lift system called "Touch Control," which incorporated dual lift cylinders for left- and right-side implements. On most Super A tractors, governors were set for 1,650 rpm, providing a substantial power boost. International Harvester produced the Super A through 1954.

The Super A was equipped with standard hydraulics, adjustable wheel-tread spacing, and an electrical system that included a starter and lights. IH produced the Farmall Super A-1 only in 1954. The A-1 was similar to the Super A, only its 122.7-ci engine turned at 1,400 rpm.

1954 Farmall Super A
The Super A was the same as the A except for the addition of the "Touch Control" hydraulic system. Farmall produced the Super A from the 1947 through 1954 model years. ***Ralph W. Sanders/ Motorbooks Archive***

The Fast-Hitch

With the unveiling of the Super C, IH introduced a unique implement-mounting concept called the Fast-Hitch, designed to counter the marketplace success of the Ford and Ferguson three-point hitch.

The Fast-Hitch featured only two points of connection. Tabs on the implement rigidly latched into receiver sockets on the two tractor lift arms. No pins were required, so the operator did not need to dismount. A feature called "Traction Control" functioned much like Ferguson's draft control by adjusting the implement's depth to keep the draft load constant. Later, when Ferguson's patents ran out and IH refused to license their system to others, the three-point system became the industry standard.

This "IH 5-Star Service" ad from 1952 features the Super C.

— THE INNOVATIVE THREE-POINT HITCH —

Henry Ford and Harry Ferguson had teamed up back in 1939 to introduce the three-point hitch on the Ford-Ferguson 9N. All went well, with the little gray tractors enjoying a marked advantage in the marketplace, until Henry Ford died in 1947. When young Henry Ford took over the company, he discovered that Ferguson was the only one making money on the arrangement. The agreement allowed either party to abrogate the deal by simply giving the other party notice. Young Henry gave such notice in 1946 and then proceeded to have the tractor upgraded to the 8N version, a 1948 model. Ferguson had begun to manufacture his own version of the tractor in England in 1946 when British Ford had elected to stick with the venerable Fordson rather than the Ford-Ferguson. When the agreement with Ford ended, Ferguson, who provided the implements and distributed the tractors in the U.S. market, began importing his English version of the tractor, the TE-20. He also began to manufacture an Americanized version, the TO-20, in the United States. Ferguson had also sued Ford for infringing upon his three-point hitch patents.

The lawsuit was settled in 1952 when Ford agreed to revise the hydraulic system on its 1953 models. The settlement also spelled out what other manufacturers had to do to avoid infringing on Ferguson's patents.

The Super C

The Super C replaced the C (which had replaced the B) in 1951. Engineers increased the engine's bore diameter by 0.125 inches, giving the Super C a displacement of 122.7 ci (up from 113.1 ci) and a 15 percent power increase. In University of Nebraska tests, the Super C was rated for two 14-inch plow bottoms. The Super C also featured ball-ramp disc brakes, a more comfortable seat, and the Touch Control hydraulic system, which was now standard rather than optional as it was on the C. Farmers could order the Super C with either a dual tricycle or wide front end.

1951 Farmall Super C
The Super C was produced from the 1951 through 1954 model years. *Ralph W. Sanders/ Motorbooks Archive*

The Super H

With the introduction of the Super H in the 1953 model year, IH designers increased engine displacement from 152.1 ci to 164 ci, boosting the horsepower to more than 30 and giving the Super H a true two-plow (16-inch) rating. The Super H also featured new disc ball ramp brakes. The distillate option, which had been offered on the Standard H, was dropped. On later models, IH engineers moved the battery to underneath the seat and added live hydraulics.

Farmall Super H
The Super had almost 30 percent more power than the H, so it was rated for two 16-inch or three 14-inch bottoms. Increasing the engine displacement from 152 ci to 164 ci gave the Super H a little more than 30 horsepower.

1953 Farmall Super H
In addition to horsepower, other improvements that were standard on the Super H included a hydraulic lift, a hydraulic snubber seat, an electrical system with starter, generator, and lights, a pressurized radiator, a swinging drawbar, a PTO and belt pulley, and disc brakes. *Ralph W. Sanders/ Motorbooks Archive*

Farmall Super H
The Super Series was offered only in model years 1953 and 1954.

1955 German Farmall DGD4

In German, the DGD letters stand for "German Big Tractor." Between 1953 and 1956, the Neuss factory manufactured 8,130 DGD4s, the German version of the Super H. The DGD4 has a coil-spring front end and a seat on the right rear fender. The DGD4 used a four-cylinder diesel of 2,175 cc (133 ci) rated at 30 horsepower.

Super M, MD, and M-TA

The Super M was introduced with the Super H in late 1952. With an engine displacement increase from 247.7 ci to 263.9 ci, the gasoline version had 22 percent more power. The Super MD, which featured a diesel engine, garnered 32 percent more power and was the first Farmall to have a 12-volt electrical system. One version of the Super M was powered by liquefied petroleum gas (LPG). It had a compression increase of between 5.9 and 6.75:1 over the gasoline version. Later in the model run, engineers moved the battery to beneath the seat and added live hydraulics.

1953 Farmall Super M
In late 1952, the Farmall Super M replaced the regular M for model year 1953. This Super M was built in Louisville, Kentucky, and has an "L" serial number. A 16.2-ci increase in engine displacement (to 263.9 ci) gave the Super M 44 horsepower on the belt at 1,450 rpm. An LPG version was added to the lineup with a compression ratio of 6.75:1 (instead of 5.9:1). The LPG version produced 46 horsepower on the belt. Otherwise, it is just like other Super Ms.

In 1954, IH introduced the Farmall Super M-TA, which featured a Torque Amplifier option—a clutch-controlled planetary auxiliary gearbox that provided a 1.482:1 ratio increase. Power-shift transmissions are quite common today, but in 1954, this was a phenomenon!

COMPARISONS

Make/Model	CID	Cyl.	Belt/PTO HP	Drawbar Pull (lb.)	Fuel Cons.*	Weight (lb.)
Super M	**263.9**	**4**	**43.92**	**5,676**	**12.01**	**8,929**
Super MD	**263.9**	**4**	**46.73**	**5,772**	**14.57**	**9,338**
Super M LPG	**263.9**	**4**	**45.66**	**6,115**	**8.76**	**9,820**
Oliver 77 LPG	293.9	6	34.49	4,778	8.46	7,510
Fordson Major D	230	4	38.49	5,315	15.5	7,892

*horsepower hours per gallon
Weight in pounds is as tested including ballast (if any) and driver.

1953 Farmall Super M
Actually made in 1952, this is the 347th Super M built. Dick Alberts of Kent City, Michigan (driving), bought the tractor from a neighborhood bar owner who bought it from a fellow down on his luck and in need of money (probably to pay his bar bill). This previous owner had been setting up the Super M as a "puller" and, accordingly, it has lots of power.

1954 Farmall Super M-TA High Clearance
This tractor is one of only sixty-four of its kind (high clearance with the wishbone wide front). The Super M-TA was the same as the Super M, except for the power-shift planetary auxiliary gearbox.

1954 Farmall Super M-TA (Super M-Torque Amplifier)

The Torque Amplifier was the first of its kind. It was a clutch-controlled planetary gearbox upstream of the regular five-speed transmission. Moving the left-hand lever provided a ratio change of 1.482:1 without disengaging the regular foot clutch. Thus, for plowing (or other heavy-duty applications), an operator would typically use a high ratio. If the tractor began to bog down, the operator could engage a low ratio by merely moving the lever. The Torque Amplifier operated much like "passing gear" in a car automatic transmission. This effectively gave the M-TA ten speeds forward and two in reverse.

After International Harvester introduced the Farmall Super M-TA, other tractor makers adopted the power-shift principle. In 1958, Ford unveiled a ten-speed all-planetary power-shift transmission. Other tractor manufacturers followed with combinations of conventional and planetary gear sets. In 1992, Agco-Allis topped all competitors when it offered a forty-eight-speed outfit with sixteen speeds in the regular gearbox plus a three-speed power-shift auxiliary.

1953 Farmall Super M-V8

This is a neat conversion of a Super M from its normal four-cylinder engine to an International truck V-8. Norm Sevcik and his son, Jeremy, did the work.

1953 Farmall Super BMD
In 1953, the International Harvester Company of Great Britain produced fifty-three BMDs in gold livery to celebrate the coronation of HRH Queen Elizabeth II. This is one of the few originals that are still in existence. The tractor and the 264-ci diesel engine were built in Britain. Alan Davis of Great Britain restored this BMD.

1954 Farmall Super MD-TA
Super Ms came in four categories, or stages. The first was the Super M of late 1952 and early 1953. Stage two of late 1953 had the battery under the seat, the LPG option, and live hydraulics. The M-TA and MD-TA were stage three, released in 1954. In stage four, Harvester renamed the Super M the Farmall 400. The Super MD-TA required quite a bit of restoration work, including a new engine, clutch, rims, tires, and some new sheet metal. Dick Alberts restored this tractor. The photo was taken in front of eighty-two-year-old Wendell Swenson's barn. Swenson, himself a collector of old Farmalls, had lived on the place for seventy-four years at the time the picture was taken in 2001.

SPECIFICATIONS

Model	M (G)	M (K)	M (D)	SM (D)	SM (G)	SM (L)	SM (D)
Years	1939-1953	1939-1953	1941-1953	1953-1954	1953-1954	1953-1954	1953-1954
Test No.	328	327	368	460	475	484	477
Test Date	Sep. 1939	Sep. 1939	June 1941	June 1951	June 1952	Sep. 1952	July 1952
Engine Cyl./CID	4/247.7	4/247.7	4/247.7	4/247.7	4/264	4/264	4/264
Rated RPM	1450	1450	1450	1450	1450	1450	1450
Drawbar HP	24.49	24.89	25.4	27.54	37.1	39.54	37.7
Belt/PTO HP	36.07	34.16	35.02	46.73	43.92	45.66	46.73
Drawbar Pull (lb.)	4,233	4,365	4,541	5,772	5,676	6,115	5,772
Test Weight (lb.)	6,770	6,770	7,570	8,555	8,929	8,950	9,338
Fuel	Gasoline	Distillate	Diesel	Diesel	Gasoline	LPG	Diesel
Transmission	5f-1r	5f-1r	5f-1r	5f-1r	5f-1r	5f-1r	5f-1r
Test Configuration	RC-Tri	RC-Tri	RC-Tri	RC-Tri	RC-Tri	RC-Tri	RC-Tri
Top Road Speed (mph)	16	16	16	16	16	16	16
Price, New ($)	2,400	2,400	2,900	3,400	3,000	3,150	3,450
Production				350,000			

1954 Farmall Super M-TA
The "Super" letter series Farmalls were only made in model years 1953 and 1954. While they were immensely popular, their short production period makes them quite collectible. Collectors have a particularly tough time finding Super M-TAs because many are still earning their keep on the farm.

1958 Farmall Super BMD
The British-built diesel tractor is virtually identical to the U.S. MD. The Super BMD has the adjustable wide front end.

1954 Farmall Super MD-TA
The Super MD had increased the engine displacement from the regular MD's 247.7 ci to 263.9 ci, giving the Super MD almost one-third more power. The Super MD was the first Farmall to have a 12-volt electrical system. The MD and the Super MD were noted for fuel economy and lugging ability. The Super MD reached a maximum engine torque of 350 foot-pounds at 1,150 rpm. The engine still produced 345 foot-pounds of torque when it had been lugged down to 1,000 rpm. Combine this strength with the Torque Amplifier (optional auxiliary gearbox) and there was not much that could bog the tractor down.

1962 Farmall B-450 Diesel
This "Made-in-England" version of the Super MD was similar to the U.S. version, but there were some differences. The British model used a glow-plug diesel, a different seat and fenders, a three-point lift, and a differential lock. The instruments were set in a separate panel box, much like the instrument panel on the big standard-tread Internationals. There is no mechanical linkage to the injectors. The "throttle" lever controls a vacuum flapper, which adjusts the injectors. The truck in the background is a 1966 Ford F-700 COE with a grain box.

SPECIFICATIONS

Model	A (G)	A (K)	B (G)	B (K)	H (G)	H (K)
Years	1939-1954	1939-1954	1939-1947	1939-1947	1939-1953	1939-1953
Test No.	329	330	331	332	333	334
Test Date	Oct. 1939	Oct. 1939	Oct. 1939	Oct. 1939	Oct. 1939	Oct. 1939
Engine Cyl./CID	4/113	4/113	4/113	4/113	4/152	4/152
Rated RPM	1400	1400	1400	1400	1650	1650
Drawbar HP	12.27	15.18	13.04	13.02	19.14	19.38
Belt/PTO HP	16.86	14.58	16.82	15.36	24.23	22.14
Drawbar Pull (lb.)	2,387	2,360	2,377	2,463	3,603	3,169
Test Weight (lb.)	3,570	3,570	3,740	3,745	5,550	5,550
Fuel	Gasoline	Distillate	Gasoline	Distillate	Gasoline	Distillate
Transmission	4f-1r	4f-1r	4f-1r	4f-1r	5f-1r	5f-1r
Test Configuration	Offset	Offset	RC-Tri	RC-Tri	RC-Tri	RC-Tri
Top Road Speed (mph)	10	10	10	10	16	16
Price, New ($)	575-1,000	575-1,000	575-1,000	575-1,000	850-1,650	850-1,650
Production	210,000				392,000	

1958 French Farmall Super FCD (French Model C Diesel)
The Super FCD was much the same as the American Super C. The French-built FD123 diesel displaced 2,016 cc (123 ci), the same as the gasoline engine in its U.S. counterpart. The diesel produced 24 horsepower on the belt. This tractor, built at the French Saint-Dizier factory, had a four-speed transmission.

1956 Farmall 300.

CHAPTER 4

Farmall by the Numbers 1954–1962

***"Not Content to Be Runner Up"* —IH Farm Equipment Division motto, 1956, taken as a challenge by John Deere**

By 1951, International Harvester's board of directors had repudiated Fowler McCormick and instated his rival, John McCaffrey, as chairman and CEO. McCaffrey instituted a less deliberate style of management, one that relied more on engineering and less on testing. McCaffrey also pushed the construction equipment product line, and some feel that the quality of the farm line suffered for it. By the late 1950s, quality problems became chronic and sales fell, leading to factory layoffs and diminishing employee morale. Under such conditions, competitors easily gained ground on IH. In 1958, IH finally relinquished its number one position in agricultural equipment to John Deere. The driving force of William Hewitt (related through marriage to the Deere family), who became chairman of Deere in 1955, may have had an impact on the standings. Shortly after assuming his post, a piece of IH dealer propaganda came across Hewitt's desk. The motto on the bottom of the page, which read, "Not Content to Be Runner Up," caught his attention. Was Deere content to be number two? Hewitt asked himself. If they were, he concluded, he would change that!

Despite losing the top spot, some remarkably beautiful and hard-working tractors emerged from the IH shops between 1954 and 1962. Harvester entered this period riding a wave of customer goodwill. In 1955, in fact, the three millionth Harvester tractor rolled off the assembly line.

An appealing new brochure sang the praises of the Farmall Cub, 100, 200, 300, and 400.

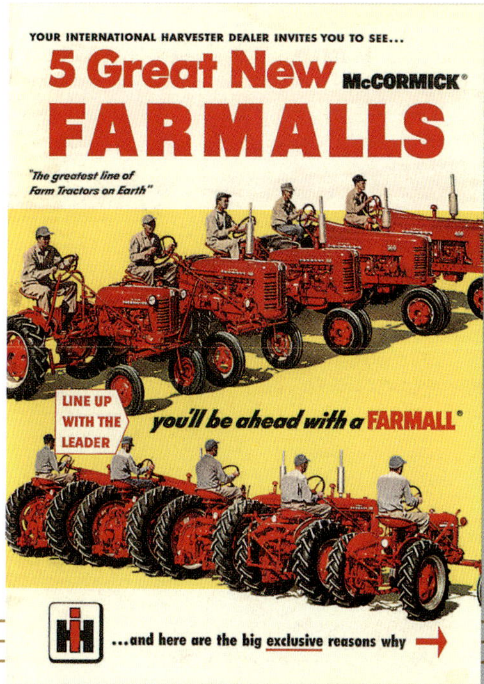

This colorful harvest scene, featuring the Farmall 400, appeared in a 1956 International Harvester calendar.

This page from a Harvester brochure introduces potential buyers to the Farmall 200, now with "Hydra-Creeper" and "Fast-Hitch."

The Farmall 200

The first of the new tractors to make the scene in 1954 (and the first Harvester model to fall under the new numbering system of identifiers) was the Farmall 200, in production through 1956. Early 200s retained the previous grille used on the Super C, while the rest of the numbered series, as well as later 200s, sported a new grille with two vertical slots and more-pronounced horizontal slots. The numbered series also featured the Farmall name in raised stainless-steel letters and numbers.

Although the Model 200 and the smaller Model 100 had the same 123-ci four-cylinder engine, operating speed for the 200 was 1,650 rpm instead of 1,400 rpm. The 200 featured special pistons and a hardened crankshaft, modifications that gave it almost 10 percent more power.

The 200 also featured an option called the Hydra-Creeper, which provided very slow operation—down to a quarter of a mile per hour—to allow for certain vegetable transplant applications. The Hydra-Creeper was essentially a hydrostatic drive. On models equipped with Hydra-Creeper, the PTO drove a hydraulic pump, which provided hydraulic fluid to a hydraulic motor that drove the transmission instead of the engine. IH later made Hydra-Creeper an option on other tractors in the line. Production of the Model 200 continued through 1956.

COMPARISONS

Make/Model	CID	Cyl.	Belt/PTO HP	Drawbar Pull (lb.)	Fuel Cons.*	Weight (lb.)
Farmall C	**113**	**4**	**19.91**	**2,902**	**11.24**	**4,409**
Farmall Super C	**122.7**	**4**	**22.9**	**3,150**	**10.78**	**5,041**
Farmall 200	**122.7**	**4**	**22.1**	**3,166**	**10.83**	**5,331**
John Deere 420	113.5	2	27.25	3,790	11.72	5,781

*horsepower hours per gallon
Weight in pounds is as tested including ballast (if any) and driver.

1955 Farmall 200
The 200, successor to the Farmall Super C, came out in 1954. The 123-ci four-cylinder engine, turning at 1,650 rpm, provided the 200 with a maximum belt horsepower of 240.11. The 200 was available from 1954 through 1956.

A group of cowboys demonstrates the abilities of the Farmall 200 in round-up style.

The Farmall 100

The Model 100 introduced in 1955 was an updated version of the A-1. The two tractors were mechanically the same, but the 100 featured a new grille and raised stainless lettering.

COMPARISONS

Make/Model	CID	Cyl.	Belt/PTO HP	Drawbar Pull (lb.)	Fuel Cons.*	Weight (lb.)
Farmall 100	**122.7**	**4**	**18.34**	**2,503**	**10.95**	**4,338**
Massey Pacer	91	4	17.88	2,118	10.54	3,469

*horsepower hours per gallon
Weight in pounds is as tested including ballast (if any) and driver.

The Farmall Cub

The Cub for 1955 was restyled with the others in the IH line, getting a new grille and raised stainless letters. Performance was improved by increasing the governor setting from 1,600 to 1,800 rpm. This increased the tractor's horsepower to 10.39 and its maximum drawbar pull to 1,605 lb. Because of the increased horsepower, the diminutive Cub was submitted to the University of Nebraska for another round of tests. Comparisons are not included here since the Cub was essentially alone in its field.

SPECIFICATIONS

Model	C	CUB	Super C	Cub (Late)	200	100
Years	1948-1951	1948-1950	1951-1954	1950-1965	1954-1956	1954-1956
Test No.	395	386	458	575	536	537
Test Date	June 1948	Sep. 1947	May 1951	June 1956	Apr. 1955	Apr. 1955
Engine Cyl./CID	4/113	4/59.5	4/122.7	4/59.5	4/122.7	4/122.7
Rated RPM	1650	1600	1600	1800	1650	1400
Drawbar HP	15	6.75	16.29	7.86	16.85	14.52
Belt/PTO HP	19.91	9.23	22.9	10.39	22.1	18.34
Drawbar Pull (lb.)	2,902	1,596	3,150	1,605	3,166	2,503
Test Weight (lb.)	4,409	2,707	5,041	2,891	5,325	4,333
Fuel	Gasoline	Gasoline	Gasoline	Gasoline	Gasoline	Gasoline
Transmission	4f-1r	3f-1r	4f-1r	3f-1r	4f-1r	4f-1r
Test Configuration	RC-Tri	RC-WF	RC-Tri	RC-WF	RC-Tri	Offset
Top Road Speed (mph)	10	6.4	10	6.4	10	10
Price, New ($)	1,300-1,500	600-800	1,500-1,850	1,000-1,750	1,900-2,200	1,400-1,800
Production	75,000	99,000	207,000	128,000	20,000	23,900

The Farmall 300

To many, the 1955 Farmall 300, with its all-red paint job and stainless-steel lettering, was the most attractive tractor ever produced by any manufacturer. Others say that as far as tractors are concerned, handsome is as handsome does. The Farmall 300 stacked up just fine under this criteria as well. And most of the approximately 30,000 Model 300s produced are still at work—testament to the tractor's quality and engineering.

The 300 replaced the Super H in 1955. New for this model were the Torque Amplifier (T/A) power-shift auxiliary transmission, a live PTO, and the Fast-Hitch two-point hydraulic lift. Engine displacement was raised to 169 ci from 164 ci by increasing the bore diameter by one-sixteenth of an inch. The operating speed was raised from 1,650 rpm to 1,750 rpm. The 300 retained the five-speed transmission, but the T/A doubled the ratios available. Speed ranged from less than 2 mph to more than 16 mph. The 300 was available in both gasoline and LPG versions, the latter having a higher compression ratio. The 300 was also available in an adjustable wide-front version and a high-clearance version. Production of the 300 continued through 1956.

1956 Farmall 300
The 300 replaced the Super H in 1955. Engine displacement increased by 5 ci and the rpm by 100 (to 1,750), giving a maximum belt horsepower of 36. The Torque Amplifier was also new on the 300. *Ralph W. Sanders/ Motorbooks Archive*

COMPARISONS

Make/Model	CID	Cyl.	Belt/PTO HP	Drawbar Pull (lb.)	Fuel Cons.*	Weight (lb.)
Farmall 300 Gas	**169**	**4**	**35.99**	**4,852**	**11.81**	**8,337**
Farmall 300 LPG	**169**	**4**	**36.74**	**4,965**	**9.31**	**8,410**
Oliver Super 66	144	6	32.83	4,309	11.38	6,623
Ford 960	172	4	44.11	4,669	11.82	6,156
John Deere 60	321	2	38.58	4,319	11.61	7,405

*horsepower hours per gallon
Weight in pounds is as tested including ballast (if any) and driver.

1956 Farmall 300
This Farmall 300 belongs to Austin Hurst of Lafayette, California. Austin's dad bought the tractor new when Austin was just a lad and then sold it in 1969. Austin went looking for the tractor when he grew up, found it, and had Dan Langy of Lena, Illinois, restore it to better-than-new condition.

The Farmall 400

The mighty 400, which rolled off IH production lines from 1955 to 1956, came with a choice of three power plants: gasoline, diesel, or LPG. The T/A power-shift auxiliary transmission was also offered, along with the Hydra-Touch hydraulic system and the Farmall Fast-Hitch. In 1956, engineers added the traction control feature. A live PTO and a belt pulley powered remote implements. Like the M and Super M before it, the 400 could not be beat for belt work. It was at about this time that hydraulic front-end loaders gained popularity, and tractors in the 400 class had the weight and brawn to make good use of one. With a loader, the optional power steering became a must.

1954 Farmall 400
The 400 replaced the Model M after just two years. As it turned out, the 400 would have a similarly brief run of just three model years, indicating the highly competitive nature of the industry at the time. The 400 could be ordered with a gasoline, diesel, or LPG engine. *Ralph W. Sanders/Motorbooks Archive*

The newly styled 400 featured a sleek hood that narrowed at the platform to improve crop visibility. The polished stainless-steel IH emblem stood out dramatically against the tractor's deep-red paint. Engineers made changes to the operator's station to improve the ergonomics, mounting instruments, an "ignition" key, and an optional tachometer in a real instrument panel.

The Farmall 400 was also available in the row-crop and high-clearance versions. International Harvester released a standard-tread variation, the International W-400, too.

(this page and photographs on following spread)
1955 Farmall 400 High Clearance
Made only in 1955 and 1956, this "Hi-Clear"400 is extremely rare. It is gasoline powered, but diesel and LPG versions were also available. The 400's regular five-speed transmission with Torque Amplifier provided ten speeds forward and two in reverse. The tractor is equipped with International's Fast-Hitch.

COMPARISONS

Make/Model	CID	Cyl.	Belt/PTO HP	Drawbar Pull (lb.)	Fuel Cons.*	Weight (lb.)
Farmall 400 Gas	**264**	**4**	**48.7**	**6,508**	**12.11**	**9,669**
Farmall 400 Diesel	**264**	**4**	**46.73**	**6,415**	**13.9**	**9,850**
Farmall 400 LPG	**264**	**4**	**49.69**	**6,374**	**9.78**	**9,579**
Oliver Super 88G	265	6	53.14	6,354	12.67	9,341
John Deere 70D	376		50.4	6,189	17.74	9,017
Massey 444 LPG	277	4	47.54	6,534	9.96	9,501

*horsepower hours per gallon
Weight in pounds is as tested including ballast (if any) and driver.

Dad moves hay with his Farmall 400 in the background of this 1956 International calendar image.

SPECIFICATIONS

Model	300 (G)	300 (L)	400 (G)	400 (D)	400 (L)
Years	1955-1957	1955-1957	1954-1956	1954-1956	1954-1956
Test No.	538	573	532	534	571
Test Date	May 1955	June 1956	Apr. 1955	Apr. 1955	May 1956
Engine Cyl./CID	4/169	4/169	4/264	4/264	4/264
Rated RPM	1,750	1,750	1,450	1,450	1,450
Drawbar HP	26.97	27.59	35.6	33.58	37.21
Belt/PTO HP	35.99	36.74	48.7	46.73	49.69
Drawbar Pull (lb.)	4,852	4,965	6,508	6,415	6,374
Test Weight (lb.)	8,337	8,410	9,669	9,850	9,579
Fuel	Gasoline	LPG	Gasoline	Diesel	LPG
Transmission	10f-2r	10f-2r	10f-2r	10f-2r	10f-2r
Test Configuration	RC-Tri	RC-Tri	RC-Tri	RC-Tri	RC-Tri
Top Road Speed (mph)	16	16	17	17	17
Price, New ($)	4,700-5,000	4,700-5,000	5,500-6,000	6,000-7,000	5,800-5,900
Production	43,000		41,000		

The Farmall 350

The Farmall 350, available from 1956 until 1958, was essentially the same tractor as the 300 but with increased engine displacement. The tractor sported a light cream-colored grille. A white "Farmall" decal decorated the sides of the hood. The stainless-steel IH emblem and numbers remained part of the design. IH introduced a diesel version that used a 193-ci Continental engine. Instead of the 169-ci engine that powered the 300, a 175-ci IH engine ran the gasoline and LPG versions.

IH offered an improved Fast-Hitch with "Pilot Guide," a feature that displayed the implement depth on an indicator the operator could easily read without turning to observe the implement.

1957 Farmall 350
Displacement on the gasoline and LPG 350s was up to 175 ci (from 169 ci on the 300, which it replaced in 1956). The Continental diesel version offered 193 ci. These placed the 350 in the just-under-40-horsepower class. The one shown has power adjustable rear wheels, power steering, Farmall Fast-Hitch, and the Torque Amplifier. Although this tractor had 3,300 hours on it when Norm Sevcik bought it, the restoration only involved new seals, new tires, and paint. To be fair, the previous owner only used the tractor for mowing and raking.

COMPARISONS

Make/Model	CID	Cyl.	Belt/PTO HP	Drawbar Pull (lb.)	Fuel Cons.*	Weight (lb.)
Farmall 350 Gas	**175**	**4**	**39.31**	**6,457**	**12.17**	**8,339**
Farmall 350 LPG	**175**	**4**	**39.34**	**6,395**	**9.60**	**8,171**
Farmall 350 Diesel	**193**	**4**	**38.65**	**6,039**	**14.79**	**8,289**
Case 511B Gas	164.5	4	44.34	5,635	11.81	7,429
Massey MF65	176	4	43.98	6,798	11.28	8,101
M-M 445	206	4	39.94	4,229	12.12	6,423

*horsepower hours per gallon
Weight in pounds is as tested including ballast (if any) and driver.

1957 Farmall 350
The cream-colored grille and side decal distinguish the Farmall 350 from its solid red predecessor, the Farmall 300. This example, restored by Dan Langy of Dan's Tractor Service in Lena, Illinois, is gas, but diesel and LPG fuel versions were also available.

This ad for the Farmall 350, which appeared in the spring 1957 issue of *Tractor Farming,* promises 20 percent more efficiency.

The Farmall 450

As a replacement for the Farmall 400, the 450 was sold between 1956 and 1958. In addition to the same cosmetic updates IH bestowed on the 350, engine displacement was increased from 264 ci to 281 ci for all versions—gasoline, LPG, and diesel. IH increased the fuel-tank capacity on gasoline and diesel versions from 18 to 20 gallons (U.S.). Owners of the Farmall 400, with the 18-gallon tank, complained of not being able to work all day on one tank of fuel as they had been able to do on their Super Ms.

COMPARISONS

Make/Model	CID	Cyl.	Belt/PTO HP	Drawbar Pull (lb.)	Fuel Cons.*	Weight (lb.)
Farmall 450 Gas	**281**	**4**	**51.55**	**7,318**	**12.47**	**8,905**
Farmall 450 Diesel	**281**	**4**	**48.78**	**6,634**	**13.57**	**9,263**
Farmall 450 LPG	**281**	**4**	**50.50**	**7,333**	**9.84**	**9,071**
Oliver 770 Diesel	216	6	48.8	7,137	14.31	9,115
Allis D-17 LPG	226	4	49.37	6,956	9.64	9,135

*horsepower hours per gallon
Weight in pounds is as tested including ballast (if any) and driver.

(this page and opposite)
1957 Farmall 450
With its 281-ci gas engine, the Farmall 450 was in the 50-horsepower class. LPG and diesel engines were also available with the same displacement. An optional wide front was also available. *Ralph W. Sanders/Motorbooks Archive*

In this ad, the Farmall 450 pulls "the world's most dependable baler," the McCormick No. 45.

1958 Farmall 450D
Farmall 450s were made between 1956 and 1958. The 450 is essentially the same as the 400, with a displacement increase from 264 to 281 ci. This 450 is the diesel version, which used a 281-ci International-built engine.

SPECIFICATIONS

Model	350 (G)	350 (L)	350 (D)	450 (G)	450 (L)	450 (D)
Years	1957-1958	1957-1958	1957-1958	1956-1958	1956-1958	1956-1958
Test No.	611	622	609	612	620	608
Test Date	Apr. 1957	May 1957	Apr. 1957	Apr. 1957	May 1957	May 1957
Engine Cyl./CID	4/175	4/175	4/193	4/281	4/281	4/281
Rated RPM	1,750	1,750	1,750	1,450	1,450	1,450
Drawbar HP	28.76	30.07	27.76	38.55	38.01	34.68
Belt/PTO HP	39.31	39.34	38.65	51.55	50.5	48.78
Drawbar Pull (lb.)	6,457	6,395	6,039	7,318	7,333	6,634
Test Weight (lb.)	8,339	8,171	8,289	8,905	9,071	9,263
Fuel	Gasoline	LPG	Diesel	Gasoline	LPG	Diesel
Transmission	10f-2r	10f-2r	10f-2r	10f-2r	10f-2r	10f-2r
Test Configuration	RC-Tri	RC-Tri	RC-Tri	RC-Tri	RC-Tri	RC-Tri
Top Road Speed (mph)	16	16	16	17	17	17
Price, New ($)	5,200	5,300	6,100	6,000	6,500	7,100
Production		28,000			39,000	

The Farmall 130

The Farmall 130, made from 1956 through 1957, was similar to the Model 100, but IH added white trim and increased the compression ratio from 6.5 to 6.94:1. The engine still operated at 1,400 rpm.

COMPARISONS

Make/Model	CID	Cyl.	Belt/PTO HP	Drawbar Pull (lb.)	Fuel Cons.*	Weight (lb.)
Farmall 130	**122.7**	**4**	**21.38**	**2,956**	**12.42**	**4,815**
Ford 640	134	4	29.46	3,008	10.99	4,200

*horsepower hours per gallon
Weight in pounds is as tested including ballast (if any) and driver.

1957 Farmall 130
The 130 was an upgrade of the Farmall 100, which superseded the Super A. The 130 was produced from 1956 until 1958. This Farmall 130 at the Sherer Maxfield dealership in Leo, Indiana, has a mounted one-bottom plow.

Featured in the spring 1957 issue of *IH Tractor Farming* magazine, this ad for the new McCormick Farmall 130 highlights the tractor's "Culti-Vision" capabilities.

The Farmall 230

Like the 130, the Farmall 230 received a compression ratio boost and a cosmetic update. It is interesting to note that the 230's performance was almost identical to the original Farmall H of 1939 but with much better fuel consumption. Especially with the Fast-Hitch, the Model 230 was a solid, two-plow tractor. The 230 was in production from 1956 until 1958.

COMPARISONS

Make/Model	CID	Cyl.	Belt/PTO HP	Drawbar Pull (lb.)	Fuel Cons.*	Weight (lb.)
Farmall 230	**122.7**	**4**	**26.71**	**3,574**	**12.16**	**5,330**
Farmall 1939 H	**152**	**4**	**24.28**	**3,603**	**11.7**	**5,500**
John Deere 420	113	2	27.25	3,790	9.98	5,781

*horsepower hours per gallon
Weight in pounds is as tested including ballast (if any) and driver.

(this and following page)
1957 Farmall 230
The Farmall 230 was in production for just three model years: 1956 through 1958. *Ralph W. Sanders/Motorbooks Archive*

New Directions in 1957

The International Harvester Board of Directors, still under the guidance of McCaffrey, elected Frank Jenks as president of the company in 1957. Jenks, a former accountant, was promoted to board chairman in 1958 when McCaffrey retired. Well aware that insufficient engineering and testing had been the practice under the former chairman and CEO, Jenks built a 442-acre research and development center near Hinsdale, Illinois, in 1959. IH hired more engineers to staff the new facility, which was called the Farm Equipment Research and Engineering Center, and delayed the introduction of new tractors for more testing.

Despite intensely tight budgets, Jenks gave his engineers some free reign to experiment with new technologies, such as turbine and free-piston engines, hydrostatic transmissions, and all-wheel drive.

The Farmall Cub, 140, and 240

Newly styled 1958 model Farmalls, including the Cub, the 140 (replacing the 130), and the 240 (replacing the 230), sported new square grilles in place of the original Raymond Loewy-styled grilles. The most important technical change that IH engineers made to these new tractors was to update the 6-volt electrical system to a 12-volt system. Other minor improvements led to the performance improvements that follow. Although the Cub featured a new governor, none of the updates required retesting at the University of Nebraska. The 240 governor was reset, increasing the engine speed from 1,650 to 2,000 rpm.

1962 Farmall 140
The 140 replaced the Farmall 130 in 1958. Production continued through 1979. Maximum belt horsepower of the C-123 engine was now up to 21. A four-speed transmission was retained. The basic configuration was still offset, just like the original Farmall A. The 140 used a 12-volt electrical system. On the Farmall 140s, live hydraulics operated the Fast-Hitch and powered remote implements. The one shown pulls a restored No. 100 IH spreader.

Although the 240 was much the same as the 230 it replaced, it looked entirely different. Rather than sitting on a platform as on the 230, the operator of the 240 sat farther forward, straddling the transmission and drive line. The steering wheel was now centered and quite flat. The U-joint arrangement to the front wheels was replaced by a drag link.

COMPARISONS

Make/Model	CID	Cyl.	Belt/PTO HP	Drawbar Pull (lb.)	Fuel Cons.*	Weight (lb.)
Farmall 140	**122.7**	**4**	**21.16**	**3,611**	**12.29**	**4,826**
Farmall 240	**122.7**	**4**	**28.32**	**4,343**	**12.01**	**6,085**
Ford 651	134	4	34.24	4,114	11.35	5,529
Oliver Super 66	139.6	4	32.83	4,309	11.38	6,623

*horsepower hours per gallon
Weight in pounds is as tested including ballast (if any) and driver.

1963 Farmall 140 Hi-Clear
This version of the popular 140 was especially handy for cultivating taller vegetables such as asparagus. Like the other 140s, this Hi-Clear has a more comfortable wide bench seat with backrest. More than 66,000 Farmall 140s, both high and regular, were produced during the tractor's twenty-one-year reign.

1960 Farmall 240
The 240 was built from 1958 through 1961 and was a more or less complete makeover of the previous model 230. It incorporated major sheet metal and modernization and featured on-center steering and seat. The C-123 engine was retained, and it produced enough horsepower to give the 240 a three-plow rating and best-in-class standing in maximum drawbar pull. The Farmall 240 gave excellent fuel economy and had a competitive price of $2,530. This one, which has seen much hard work, belongs to Norval Poyser, shown at the controls.

Farmall 340

IH unveiled the predecessor to the Farmall 340—a new International utility tractor called the IH 330—in the spring of 1957. Although never made as a Farmall, they combined elements of the International 350 with a new four-cylinder gasoline engine of 135 ci, which IH engineers had designed for the forthcoming Farmall and International 340. Since those new tractors were not yet ready for release, engineers developed the International 350 by mating the new 2,000-rpm engine to the I-350 chassis. IH hoped that the introduction of the 330 would fill a void in the line and enable the company to test the engine in the meantime.

When the 340 rolled off the production line in late 1957, however, it was a totally new design. Resembling an upscale 240, the tractor was available in a variety of configurations, including an International Utility model, a grove model, and the T340 crawler. In addition to the 135-ci gasoline engine, IH also produced a new diesel of 166 ci and 2,000 rpm.

The 340 featured a new hydraulic system, which used transmission oil as the hydraulic fluid. Filters and valves were in the transmission case. Pumps, either in the transmission housing or on the engine, the two-point Fast-Hitch, and the three-point (Ferguson-type) hitch were optional. The 340 featured a new, flat instrument panel and T/A power shift. Power steering was only offered on the International version. More than 7,000 Farmall 340s were built between 1958 and 1963.

COMPARISONS

Make/Model	CID	Cyl.	Belt/PTO HP	Drawbar Pull (lb.)	Fuel Cons.*	Weight (lb.)
Farmall 340 Gas	**135**	**4**	**32.3**	**5,142**	**12.45**	**6,761**
Farmall 340 Diesel	**166**	**4**	**38.93**	**36.14hp†**	**13.81**	**7,579**
Case 441 Gas	148	4	33.11	29.15hp†	11.85	6,202
Massey MF35D	152.7	3	37.04	32.13hp†	15.69	5,909

*horsepower hours per gallon
Weight in pounds is as tested including ballast (if any) and driver.
†After February 1959, changes in Nebraska tractor test procedures reported maximum power drawbar pull in horsepower. To achieve the best results, manfacturers opted for a lesser pull in pounds in order to gain more speed and less slippage.

1959 Farmall 340
Hal Beitlich of Rockford, Illinois, was photographed at the controls of his Farmall 340. Hal was a farm boy from around La Crosse, Wisconsin, before going to the University of Wisconsin and becoming a contracts administrator for a large corporation. The 340 helped him remember his roots. Roger Blumer of Monroe, Wisconsin, did the renovation.

1959 Farmall 340
The 340, unveiled in 1958, was a totally new design. It was available in gasoline or diesel versions (gasoline shown here). The gasoline engine was the C-135 (135-ci) four-cylinder with overhead valves. Transmission lubricant was used as hydraulic fluid. An engine-driven (live) hydraulic pump was an option. The 340, a very popular and capable tractor, was built from 1958 through 1963. This beauty has the Farmall Fast-Hitch, although a three-point hitch was an option on the 340.

1961 Farmall 340 Diesel
Although both the gasoline and the diesel version of this tractor had four-cylinder engines of the OHV type, the 340's new diesel engine displaced 166 ci compared to the gasoline version's 135 ci. On the 1961 Farmall 340, a rather complex system of shafts and universal joints connected the steering wheel to the front wheels (the same type of linkage was used on the 240). The system worked well with the on-center steering wheel placed at a comfortable angle.

SPECIFICATIONS

Model	130	230	140	240	340 (G)	340 (D)
Years	1956-1957	1956-1958	1958-1975	1958-1961	1958-1963	1958-1963
Test No.	617	616	666	667	665	775
Test Date	May 1957	May 1957	Sep. 1958	Sep. 1958	Sep. 1958	Nov. 1960
Engine Cyl./CID	4/122.7	4/122.7	4/122.7	4/122.7	4/135	4/166
Rated RPM	1,400	1,800	1,400	2,000	2,000	2,000
Drawbar HP	16.06	20.05	16.41	21.55	24.87	27.95
Belt/PTO HP	21.38	26.71	21.16	28.32	32.3	38.93
Drawbar Pull (lb.)	2,956	3,574	3,611	4,343	5,142	5,559
Test Weight (lb.)	4,815	5,330	4,826	6,085	6,761	7,579
Fuel	Gasoline	Gasoline	Gasoline	Gasoline	Gasoline	Diesel
Transmission	4f-1r	4f-1r	4f-1r	4f-1r	10f-2r	10f-2r
Test Configuration	Offset	RC-Tri	Offset	RC-Tri	RC-Tri	Utility
Top Road Speed (mph)	10	12	13	15	17	17
Price, New ($)	2,000	2,200	2,400-4,300	2,300-3,000	3,600	4,300
Production	15,000	12,000	60,000	4,200	8,000	

The Farmall 560

In the summer of 1958, International Harvester unveiled its new line of big tractors. The first to make the scene was the six-cylinder Model 560, built from 1958 until 1963. One might say that it supplanted the Farmall 450, but there was really no relation between the two. The new tractor was available with gasoline, LPG, or diesel power plant. The diesel had greater piston displacement at 283 ci over the other two at 263 ci. The 560 was touted as the world's most powerful row-crop tractor.

After 1958, IH abandoned the switchover starting system for its diesel engines. The new D-282 engine relied on glow plugs. Later, IH offered a turbocharger for the 560 diesel, which was not intended to raise the horsepower but to improve fuel consumption and maintain rated horsepower at higher altitudes.

The Farmall 560 is beautifully rendered in this 1971 International Harvester calendar image.

COMPARISONS

Make/Model	CID	Cyl.	Belt/PTO HP	Drawbar Pull (lb.)	Fuel Cons.*	Weight (lb.)
Farmall 560G	**263**	**6**	**61.04**	**7,588**	**12.65**	**9,239**
Farmall 560L	**263**	**6**	**57.06**	**7,524**	**9.73**	**9,361**
Farmall 560D	**282**	**6**	**58.48**	**7,347**	**14.46**	**9,461**
Cockshutt 570D	298	6	60.84	9,063	15.03	11,335
Oliver 880G	265	6	57.43	7,998	12.20	10,799
Massey MF 85G	242	4	61.23	N/A	11.64	11,245

*horsepower hours per gallon
Weight in pounds is as tested including ballast (if any) and driver.

1959 Farmall 560D
The big Farmall 560 made its debut on July 18, 1958. Available in gasoline, LPG, or, as seen here, diesel, the 560 was the first of many six-cylinder Farmalls. The first two variants displaced 263 ci and the diesel, 282 ci. By the time the diesel 560 rolled off the production lines, Harvester had abandoned the complex but effective switchover starting method in favor of glow-plug starting. All three engines gave the 560 a five-plow rating. *Ralph W. Sanders/Motorbooks Archive*

1961 Farmall 560 Demo
Power steering and the wide seat with backrest were now standard equipment on the 560. Optional features included the Torque Amplifier, a faster first gear and reverse (excellent for loader work), and the Fast-Hitch. The Farmall 560 shown here is a Brass Tacks Demonstrator. A special package of factory- or dealer-installed decals, which pointed out the features of the new tractor, was available.

1960 Farmall 560 Puller
DuWayne Goings of Auburn, Indiana, is into the antique tractor pulling hobby (1960 and older) in a big way. This 560 gas-powered model performs well even in the 10,000-lb. class (with additional weights on the rear wheels).

The Farmall 460

The four-plow Farmall 460 had all the features of the five-plow 560 but was smaller. Its six-cylinder gasoline and LPG engines displaced 221 ci. The diesel displaced 236 ci. The over-engine steering that had characterized the big Farmalls was now more like that of the old Farmall C, except the steering wheel and operator were in the center. The steering shaft, which connected the steering wheel to the front wheels through universal joints, ran along the left side of the engine. Power steering was standard on both the 460 and 560, which had the same steering arrangement. Both also offered the T/A power shift and 12-volt electricals. IH manufactured the 460 from 1958 through 1963.

COMPARISONS

Make/Model	CID	Cyl.	Belt/PTO HP	Drawbar Pull (lb.)	Fuel Cons.*	Weight (lb.)
Farmall 460G	**221**	**6**	**46.94**	**6,734**	**12.23**	**8,931**
Farmall 460L	**221**	**6**	**46.86**	**6,903**	**9.29**	**9,316**
Farmall 460D	**236**	**6**	**50.1**	**6,846**	**14.04**	**9,151**
Massey MF65G	176	4	43.98	6,798	11.28	9,057
Oliver 770D	216	6	48.8	7,137	14.31	9,115

*horsepower hours per gallon
Weight in pounds is as tested including ballast (if any) and driver.

1963 Farmall 460
This 460 used a six-cylinder engine with 221 ci for the gasoline and LPG versions and 236 ci for the diesel. Otherwise, its configuration and options matched those offered on the 560, including the draft-control three-point hitch. In addition to the narrow- and wide-front Farmall versions, IH also produced International Utility versions, Wheatland versions, and a Grove model for orchard work. Only three Grove models are known to exist.

1963 Farmall 460
International Harvester manufactured the 460 from 1958 through 1963. Bob Daws of Lena, Illinois, bought this 460 wide-front new in September 1963. He used it with his corn planter for twenty-seven consecutive years before he retired it.

Bad Days at Harvester, 1958-1963

International Harvester had the knack, as most of the comparison figures indicate, of getting more out of their tractor models, especially in the area of maximum drawbar pull. Because these tractors were generally loaded heavier than the competition, they got more pull. Unfortunately, the additional pull was sometimes more than the gears

SPECIFICATIONS

Model	560 (G)	560 (L)	560 (D)	460 (G)	460 (L)	460 (D)
Years	1958-1963	1958-1963	1958-1963	1958-1963	1958-1963	1958-1963
Test No.	671	675	669	670	676	672
Test Date	Oct. 1958	Oct. 1958	Sep. 1958	Oct. 1958	Oct. 1958	Oct. 1958
Engine Cyl./CID	6/263	6/263	6/282	6/221	6/221	6/236
Rated RPM	1,800	1,800	1,800	1,800	1,800	1,800
Drawbar HP	44.83	43.96	44.38	35.95	36.17	36.19
Belt/PTO HP	61.04	57.06	58.48	46.94	46.86	50.7
Drawbar Pull (lb.)	7,588	7,524	7,347	6,734	6,903	6,846
Test Weight (lb.)	9,239	9,361	9,461	8,931	9,316	9,151
Fuel	Gasoline	LPG	Diesel	Gasoline	LPG	Diesel
Transmission	10f-2r	10f-2r	10f-2r	10f-2r	10f-2r	10f-2r
Test Configuration	RC-Tri	RC-Tri	RC-Tri	RC-Tri	RC-Tri	RC-Tri
Top Road Speed (mph)	17	17	17	17	17	17
Price, New ($)	6,000	6,100	6,700	4,700	4,800	5,400
Production		70,000			35,000	

could take. This problem, plus McCaffrey's policy of sparing lengthy product testing, led to a catastrophe of gigantic proportions. After about 300 hours of normal to heavy usage, the final drives of the 560 and the 460 began to fail in the field.

Despite cash flow problems, IH engineers worked like crazy to fix the problem, revising tractors on the line as they worked out the details. Eventually, they redesigned the entire differential case. To restore farmer confidence, IH doubled warranties and repaired failed tractors as quickly as possible. In many cases, IH provided loaner tractors to keep the farmer going.

Over the first few years in operation, there were more problems with these tractors, including crankshaft breakage, clutch failure, and steering problems, but IH stepped up to them all. While the 460 and 560 are among the most reliable machines today, this string of events allowed an aggressive John Deere to pass IH in production.

The Farmall 404

The 404, introduced as a 1961 model in 1960, was the first product of the new Hinsdale, Illinois, Farm Equipment Research and Engineering Center. This 442-acre campus was built to correct some of the neglect of proper engineering and testing that had characterized IH in the past. The new 404 featured a 135-ci four-cylinder gasoline- or LPG-fueled engine and a four-speed transmission with the two-range auxiliary gearbox, providing eight total speeds forward and two in reverse. Although it was mostly a new design, the 404 used the 330/340 gasoline-LPG engine. It marked Harvester's move to a Ferguson-type three-point hitch, which by then had become the industry standard. Power steering was totally hydrostatic, and the 404 featured a dry air filter, both firsts that would also become the industry norm.

The first 404s had ground-driven (only operated with the main clutch engaged) PTOs—a live PTO was not an option. Bitter complaints from dealers that sales were going to the competition because of this so-called cost-saver finally reached IH management. IH engineers designed and fitted a live PTO for production in about 1964. The 404 was built in model years 1961 through 1967.

COMPARISONS

Make/Model	CID	Cyl.	Belt/PTO HP	Drawbar Pull (hp)	Fuel Cons.*	Weight (lb.)
Farmall 404G	**134.8**	**4**	**36.7**	**32.97**	**12.34**	**6,645**
Farmall 404L	**134.8**	**4**	**(No Test)**			
Massey MF505D	152.7	3	38.33	32.42	15.72	6,017
Allis D-12	149	4	33.32	29.43	12.21	5,630

*horsepower hours per gallon
Weight in pounds is as tested including ballast (if any) and driver.

1962 Farmall 404
All new when it came out in late 1960 as a 1961 model, the 404 featured full hydrostatic steering with no mechanical connection between the front wheels and the steering wheel. The 404 was also the first Harvester tractor to have the draft-control three-point hitch as standard. The four-cylinder OHV engine, available in gasoline and LPG versions only, displaced 135 ci.

The Farmall 504

The 504 incorporated more new features than the 404. The most startling to IH buyers (although the feature had previously appeared on John Deere tractors) was the fully hydrostatic power steering, which operated with no mechanical connection between the steering wheel and the tractor's front wheels. The engine-driven hydraulic pump provided the steering muscle. The 504 was available in gasoline, LPG, and diesel versions. The gas and LPG engines displaced 152.1 ci, while the diesel displaced 187.5 ci. Like the 404, the 504 featured a draft-control three-point hitch and a dry-type air filter. Unique to the 504 was a hydraulic oil cooler mounted in front of the radiator, a live PTO, and a five-speed plus T/A transmission.

In response to cane-growers' demand for a tractor, IH introduced a high-clearance version of the 504 in 1963. The 504 was in production from 1961 to 1968.

1962 Farmall 504 Diesel
The 504's four-cylinder OHV engine, which produced about 45 horsepower, was available in gasoline, LPG, or diesel versions. Gasoline and LPG models displaced 152 ci, while the diesel displaced 187 ci. This 504 Diesel retained the same features as the 404, including full hydrostatic steering and the draft-control three-point hitch.

COMPARISONS

Make/Model	CID	Cyl.	Belt/PTO HP	Drawbar Pull (hp)	Fuel Cons.*	Weight (lb.)
Farmall 504G	**152.1**	**4**	**46.2**	**40.41**	**12.33**	**7,870**
Farmall 504LPG	**152.1**	**4**	**44.36**	**41.21**	**8.99**	**7,895**
Farmall 504D	**187.5**	**4**	**45.99**	**38.9**	**14.89**	**8,075**
John Deere 2010G	144.5	4	46.86	39.12	11.79	6,310
Allis D-15G	149	4	40.0	35.33	11.86	7,481

*horsepower hours per gallon
Weight in pounds is as tested including ballast (if any) and driver.

SPECIFICATIONS

Model	404 (G)	504 (G)	504 (L)	504 (D)
Years	1961-1967	1961-1968	1961-1968	1961-1968
Test No.	818	819	820	816
Test Date	June 1962	June 1962	June 1962	May 1962
Engine Cyl./CID	4/134.8	4/152.1	4/152.1	4/187.5
Rated RPM	2,000	2,200	2,200	2,200
Drawbar HP	27.5	32.7	32.91	31.95
Belt/PTO HP	36.7	46.2	44.36	45.99
Drawbar Pull (lb.)	4,999	6,111	6,178	6,270
Test Weight (lb.)	6,645	7,870	7,895	8,075
Fuel	Gasoline	Gasoline	LPG	Diesel
Transmission	8f-2r	10f-2r	10f-2r	10f-2r
Test Configuration	RC-WF	RC-Tri	RC-Tri	RC-Tri
Top Road Speed (mph)	21	18	18	18
Price, New ($)	3,300	3,800-4,300	3,900-4,400	4,400-4,800
Production	3,000	————	17,000	————

1965 Farmall 806.

CHAPTER 5

Farmalls of a New Generation 1963–1973

"Nothing recedes like success."
—Walter Winchell, 1954

In the early 1960s, life in the United States was in a state of flux. When President Kennedy was assassinated on November 22, 1963, in Dallas, Texas, and Vice President Lyndon Johnson took over as president, the growing divide over civil rights and the escalating war in Vietnam threatened to tear the country apart.

Meanwhile, more and more of America's small acreage farmers were selling to neighbors who had the desire to expand. Oftentimes, older farmers sold their land when they retired because their children were not interested in doing the hard work for the small payoff. Larger-scale operations, on the other hand, could be profitable, and the new machinery available allowed the farmer to handle more acres without hired help. With more land to cultivate and no more hours in the day, farmers had to work at a faster rate, and in order to do that, they needed more horsepower.

In 1960, the average farm tractor had 40 to 50 horsepower. Prior to 1960, the Oliver 995 provided the most power in a two-wheel-drive farm tractor at 85 horsepower, followed by the International 660 at 82 horsepower, and the John Deere 830 at 73 horsepower. In the early 1950s, the Funk brothers, foundry men and aircraft builders from Coffeeville,

A brochure cover accentuates the grille and hood design on what would be the last generation of Farmalls.

The Farmall 806 surges ahead in this image from a 1971 International Harvester calendar.

Kansas, converted some 5,000 Ford N Series tractors to Ford 6 and V-8 engines, increasing their horsepower up to 85!

At the beginning of the 1960s, the big, four-wheel-drive tractors were waiting in the wings. In 1958, the Steiger brothers assembled their first four-wheel-drive articulated tractor from earthmover parts in their Minnesota barn. A 238-horsepower Cummins engine powered the tractor. John Deere released its 215-horsepower 8010 articulated 4WD in 1959. Interestingly, the John Deere 8010, powered by a six-cylinder engine, was the first Deere with more than two cylinders since the ill-fated Dain-John Deere of 1917. To meet Deere's challenge, Harvester turned to its Hough Industrial Division, which built earthmovers for the company. The Hough-built Model 4300 four-wheel drive, unveiled in 1960, was an awesome machine with an 817-ci turbocharged diesel of 300 horsepower, weighing almost fifteen tons. Neither it nor the 8010 were well received.

The inner workings of a 1967 Farmall 806

The four-number designator of the John Deere 8010 indicated that something new was definitely in the works at Deere. On August 29, 1960, Deere unveiled its "New Generation of Power" line of tractors, all with three-, four-, and six-cylinder engines and four-number model designators. The six-cylinder 4010 was the most astounding tractor of the line, featuring a central aircraft-type hydraulic system that operated the implement lift, brakes, and steering.

— THE WORLD TRACTOR CONCEPT —

In the early 1960s, leading tractor manufacturers began adopting standardized tractor models to sell in all the countries that they serviced. Factories in the various countries would each make a model or implement of the total product line, shipping them to dealers around the world. Styling and features would be as consistent as possible. John Deere, for example, had factories in Spain, Germany, Argentina, France, Mexico, and South Africa, as well as those in the United States and Canada. In 1964, Ford combined its British Fordson division with its Dearborn, Michigan, division and its German operations with its Canadian operations. Massey-Ferguson (then the world leader in tractor unit sales) had been "international" for years. International Harvester initiated its "World Wheel Tractor" series in 1967. At first, IH was somewhat slow to adopt the World Tractor concept. It was Brooks McCormick (great-grandson of William and Cyrus Senior's youngest brother), elected chairman and CEO in 1968, who saw the World Tractor concept as a way to streamline operations, cut costs, and make more money on the tractor line.

Harvester engineers were not totally unprepared, however. In early 1959, Harvester initiated a total redesign of drive lines to meet the rapidly increasing horsepower demands of the farming industry. When IH engineers got wind of the new John Deere 4010, they knew they had their marching orders. Previous final drives had been based on the Farmall M of 1939. This drive line stretching was basically the source of the problems encountered with the 460 and 560 tractors that had proven to be so costly to IH.

The Farmall 806

The Farmall 806, one of the greatest tractors of all time, carried the fight to John Deere. Like the JD 4010, the 806 sported the new central hydraulic system, but instead of a single circuit for all functions, the 806 had three—one for steering, one for brakes, and one for the implement lifts. Since neither the JD 4010 nor the IH 806 had mechanical linkage to steer the front wheels, a hose blowout on the Deere could leave the operator with only marginal steering or brakes.

The Farmall 806 was billed as the world's most powerful all-purpose tractor. Its six-cylinder 361-ci diesel produced 94.93 horsepower during tests at the University of Nebraska. Deere's 4020 (which by then had replaced the 4010) mustered a mere 91.17 horsepower. In addition to the diesel, the Farmall 806 offered a six-cylinder 301-ci gasoline or LPG engine, producing about the same power. But, the tractor was configured to accept the taller diesel engine, so when the shorter gasoline/LPG engine was installed, too much daylight shone between the top of the engine and the hood side panels. This, IH officials reasoned, was too reminiscent of the two-cylinder Deere. So IH engineers devised a cast-iron valve cover that was tall enough to fill the gap.

The new 806 diesel was of the direct-injection type. To assist in cold-weather starting, IH pioneered an ether injection system. Originally, the operator used a cable that triggered the valve on a replaceable can to direct ether into the intake manifold while the engine was cranking. After reports of the cable freezing, IH installed a push-button-operated electric solenoid valve instead. IH offered kits to owners who wanted to replace the cable with the electric valve start.

1967 Farmall 806
Built from 1963 until 1967, the 806 was billed as the world's most powerful tricycle tractor. Many enthusiasts go so far as to say that the 806 was the best tractor ever produced! The success of the 806 gave International Harvester some reprieve after the 460/560 drivetrain debacle of 1959. The 806 featured new hydrostatic steering (no mechanical connection), a new hydraulically actuated Torque Amplifier, and hydrostatic power brakes. Although the 806 was available in gasoline, LPG, and diesel versions, the German-built, direct-injection diesel was by far the most popular.

The 806 Farmall transmission controls were mounted on the right side of the steering "tower." There were two levers: The inner lever moved fore and aft in a straight line to select high or low range, and reverse or neutral. A second lever selected one of four main gear ratios for eight speeds forward or four in reverse. With the optional two-speed Torque Amplifier, controlled with a lever on the left side of the tower, the 806 had sixteen forward speeds and eight in reverse, offering enough flexibility for almost any occasion, it would seem. The new Deere 4020, however, offered eight speeds forward, four speeds in reverse, shift-on-the-go, and no-clutch one-lever shifting. Because of its ease of use, large-scale farmers who employed hired men to drive tractors often opted for the Deere. The novice driver found it much easier to handle the Deere power-shift transmission.

The Farmall 806 incorporated two live PTO shafts—one operating at 540 rpm and the other at 1,000 rpm. For the most part, flat belt pulleys disappeared from tractors of fifty or more horsepower by 1960. A lever behind the seat activated a hydraulic clutch, which controlled the PTOs. The clutch had its own hydraulic pump. IH included a brake as a safety feature to stop the PTO shafts instantly when disengaged.

Aside from the three engine choices, IH offered the 806 in a variety of configurations. The farmer had a choice of narrow or wide front ends, a fixed or swinging drawbar, the two-point Fast-Hitch or three-point hitch. A high-clearance version and a

1967 Farmall 806
In its Nebraska test, the 806 achieved almost 95 horsepower. Its six-cylinder diesel displaced 361 ci, while gas and LPG versions displaced 301 ci. The 806 was available with either a narrow or wide front end. This one has a nice IH retrofit cab.

1965 Farmall 806
Ed Engen of Brodhead, Wisconsin, uses this gasoline version of the 806 as a "puller" in the 9,000–10,000-lb. class. Ed, who also pulls with a '66 Cub Cadet, bought the 806 new.

factory-installed Front Wheel Assist (FWA) front axle were also options. And, in January 1965, IH made a factory cab available.

Between 1963 and 1967, the time during which the 806 was built, almost 43,000 Farmall 806s left the production line, and most of the machines are still soldiering on. Reports indicate that many of the diesel engines reach more than 30,000 hours before they require an overhaul.

COMPARISONS

Make/Model	CID	Cyl.	Belt/PTO HP	Drawbar Pull (hp)	Fuel Cons.*	Weight (lb.)
Farmall 806D	**361**	**6**	**94.93**	**84.77**	**14.96**	**11,895**
Farmall 806G	**301**	**6**	**93.27**	**80.70**	**12.58**	**11,045**
Farmall 806L	**301**	**6**	**93.42**	**81.53**	**9.72**	**11,525**
John Deere 4010	380	6	84.00	71.93	14.97	9,775
John Deere 4020	404	6	91.17	76.36	14.2	13,055

*horsepower hours per gallon
Weight in pounds is as tested including ballast (if any) and driver.

The Farmall 706

The 706 model Farmall was a slightly reduced scale version of the 806. Its six-cylinder diesel displaced 282 ci, while the gasoline/LPG six displaced 263 ci. The diesel was glow-plug start, rather than a direct injection as on the 806.

The 706 featured all of the options and versions offered on the 806, including the FWA package, making the 706 Farmall even more popular than its big brother. Production of the 706 ceased in 1967.

COMPARISONS

Make/Model	CID	Cyl.	Belt/PTO HP	Drawbar Pull (hp)	Fuel Cons.*	Weight (lb.)
Farmall 706G	**263**	**6**	**73.82**	**66.18**	**12.86**	**9,895**
Farmall 706D	**282**	**6**	**72.42**	**65.12**	**13.15**	**10,011**
Farmall 706L	**263**	**6**	**73.66**	**66.06**	**9.88**	**9,950**
John Deere 3020D	270	4	65.28	54.77	12.71	9,585

*horsepower hours per gallon
Weight in pounds is as tested including ballast (if any) and driver.

1966 Farmall 706
Aaron Woker of Pearl City, Illinois, drives his gasoline-powered 706 Farmall. The 706 was produced from 1963 through 1967. The 706's original six-cylinder gasoline and LPG engines, which powered these tractors from 1963 to 1966, displaced 263 ci, producing a maximum belt horsepower of 74. In late 1966, International Harvester increased the displacement to 291 ci for a maximum horsepower of 77. With this increase, gasoline and LPG tractors could keep pace with the diesel version, which at that time changed to a German-built direct-injection engine of 310 ci, up from 282 ci in the original version.

A family examines the features on the new Farmall 706 on display at the fair.

1963 Farmall 706 Diesel
The Farmall 706 was much the same as its big brother the 806, only it was a little smaller. It was very popular with farmers, and more than 52,000 were delivered. It was produced from 1963 to 1967.

The Farmall 1206

Another great tractor from International Harvester was the 100+ horsepower Farmall Model 1206. The 1206 was the first of the Farmall line to carry a turbocharger. The unit, which used an exhaust gas turbine to drive a compressor, was built by Solar, a San Diego company that IH had acquired a few years earlier to take advantage of its gas turbine engine technology. A few years later, when business began to unravel at Harvester, Solar was sold to Caterpillar.

IH engineers redesigned and strengthened the basic 361-ci diesel used in the 806 for use in the 1206 turbo, employing new lubrication schemes, using a hardened seven-main crank, adding a larger air cleaner and cooling fan, and extensively strengthening the drive train. During its Nebraska test, the engine exceeded 113 horsepower through the PTO.

The 1206 featured an eight-speed manual transmission with an optional Torque Amplifier. A factory cab, wide or narrow fronts, and factory dual rear wheels were also options. The 1206 was only available with the 1,000 rpm PTO, as torque levels using the slower PTO were just too high for drive shafts.

On the 1206, IH engineers incorporated new flat-top fenders, similar to those John Deere had used on their tractors for about eight years. The new fenders held the headlights and taillights and had handholds for the operator while stepping up to the platform. Flat-topped fenders were subsequently added as options or replacements on the 706 and 806 Farmalls.

The original 1206 was painted and styled like the 706 and 806 Farmalls, but a welded, tubular grille replaced the cast units of the earlier tractors. A "1206 Diesel" decal appeared on each side of the hood. On later model 1206s, Harvester painted the grille, fenders, and wheels white and added a gold-script "Turbo" decal to each side of the hood. In 1966, IH replaced the 1206 decals with metal emblems.

COMPARISONS

Make/Model	CID	Cyl.	Belt/PTO HP	Drawbar Pull (hp)	Fuel Cons.*	Weight (lb.)
Farmall 1206TD	**361**	**6**	**112.64**	**95.00**	**15.95**	**13,580**
John Deere 5020D	531	6	133.25	113.72	16.21	21,360
Massey 1130D	354	6	120.51	106.91	15.86	18,460

*horsepower hours per gallon
Weight in pounds is as tested including ballast (if any) and driver.

1967 Farmall 1206 Diesel
The turbocharged Farmall 1206 Diesel was the first row-crop tractor with more than 100 horsepower available to the farmer. It produced almost 113 horsepower during its Nebraska test. Harvester's Solar Division manufactured the turbocharger. Harvester used white trim to make the 1206 stand out. The Farmall 1206 was built from 1966 to 1967.

SPECIFICATIONS

Model	806 (D)	806 (G)	806 (L)	706 (D)	706 (G)	706 (L)	1206 TD
Years	1963-1967	1963-1967	1963-1967	1963-1967	1963-1967	1963-1967	1966-1967
Test No.	857	859	861	856	858	860	910
Test Date	Oct. 1963	Nov. 1963	Nov. 1963	Nov. 1963	Nov. 1963	Nov. 1963	Sep. 1965
Engine Cyl./CID	6/361	6/301	6/301	6/282	6/263	6/263	6/361
Rated RPM	2,400	2,400	2,400	2,300	2,300	2,300	2,400
Drawbar HP	84.77	80.7	81.53	65.12	66.18	66.06	95
Belt/PTO HP	94.93	93.27	93.42	72.42	73.82	73.66	113
Drawbar Pull (lb.)	9,182	9,003	9,072	8,026	7,976	8,149	10,744
Test Weight (lb.)	11,895	11,045	11,525	10,011	9,895	9,950	15,583
Fuel	Diesel	Gasoline	LPG	Diesel	Gasoline	LPG	Diesel
Transmission	16f-4r	16f-4r	16f-4r	16f-4r	19f-4r	19f-4r	16f-4r
Test Configuration	RC-Tri	RC-Tri	RC-Tri	RC-Tri	RC-Tri	RC-Tri	RC-WF
Top Road Speed (mph)	20	20	20	18	18	18	20
Price, New ($)	7,900	7,100	7,200	7,100	6,300	6,400	9,500
Production		35,000			46,000		20,000

The Farmall 656

An International 606, replacement for the 460, was released with the 806 and 706 Farmalls. A little more than two years later, a revised version of the 606 was produced as both the Farmall 656 and the International 656. The Farmall 656 was available in either a gasoline version, which displaced 263 ci, or a diesel engine version, which displaced 281 ci. Both were six-cylinder engines producing around 60 PTO horsepower. A five-speed transmission with the Torque Amplifier was featured.

The 656 and other "56 Series" tractors were the first to feature the two-post Roll-Over Protection System (ROPS) for tractors without cabs. A sun-rain shade was available with the ROPS.

In 1967, the Farmall 656 was equipped with a fully hydrostatic transmission, enabling operators to obtain an infinite range of gear ratios by simply moving the S/R

(Speed Ratio) lever. In tractors with a hydrostatic transmission, the engine drives a carriable displacement hydraulic pump, which in turn supplies hydraulic flow to a fixed or variable displacement hydraulic motor connected to the final drive.

Sundstrand Aviation helped IH develop the basic hydrostatic transmission for the turbine-powered International HT-341 concept tractor in 1961. Sundstrand had perfected the continuously variable hydrostatic drive long before World War II as a way to change the feed rate of its machine tools. During the war, gun turret drives for bombers used hydrostatic technology. After the war, increasing electric power demands in aircraft prompted Sundstrand to develop hydrostatic alternator drives (such alternators must run at an exact, constant speed while the engine shifts from idle to full speed). Sundstrand also developed hydrostatic transmissions for a variety of applications from garden tractors to highway trucks.

1967 Farmall 656 Hydro
The first hydrostatic drive tractor available to the farmer was the 60-plus-horsepower Farmall 656 Hydro. A single lever controlled the displacement of the engine-driven hydraulic pump, which in turn drove hydraulic wheel motors. Changing the pump displacement effectively changed the gear ratio of the drive, passing through zero displacement for stopping, and in the "negative direction" for reverse. Engine speed could remain constant through this cycle. This feature was especially helpful when a portion of the power was used through the PTO and the operator wanted to keep a steady engine speed.

The use of the hydrostatic transmission in the 656 tractor seemed to be an ideal application. In addition, the 656 included a two-range auxiliary to provide full-rpm speeds from barely creeping to 21 mph. On the downside, straight hydrostatic transmissions were only 70 to 80 percent efficient, whereas gear transmissions are generally 98 percent efficient. That translates into lower drawbar power and large quantities of heat to dissipate with the hydrostatic.

— SOLAR AIRCRAFT COMPANY —

The Solar Aircraft Company of San Diego, California, was founded just before World War II. At a time when the technology for welding such high-temperature components was a rare commodity, Solar made a name for itself by fabricating exhaust manifolds for aircraft engines. After the war, as the demand for large piston aircraft engines began to diminish, Solar used the metallurgical technology to develop simple gas turbine engines. These turbines were used for ship propulsion, pipeline pumping, powering of onsite electrical generating plants, and for aircraft auxiliary power units (APUs).

The U.S. Defense Department was Solar's best customer until 1957, when the Soviets launched *Sputnik I* into the earth's orbit, and Charles "Engine Charlie" Wilson (Eisenhower's defense secretary) notified defense contactors that military spending would shift from aircraft to missiles. Solar's sales dropped by almost 70 percent, making the company ripe for a corporate takeover. Herbert Kunzel, Solar's president, approached Harry Bercher, then executive VP of Harvester, about the possibility of incorporating Solar turbine engines into Harvester's line. Harvester, which had already been experimenting with a Boeing-built gas turbine as a propulsion unit for trucks and heavy equipment, quickly agreed to buy Solar in 1957.

The engineers at Solar developed experimental engines for trucks and tractors. Then, in the early 1970s, the Arab oil embargo tripled the price of fuel. Higher fuel prices coupled with higher fuel consumption of turbines made diesel engines the more logical choice for trucks and tractors. So, in 1981, at a time when the aircraft and oil field markets had rebounded to the point that Solar was producing a third of International Harvester's profits, IH sold the company to Caterpillar to get a cash inflow to offset IH's mounting financial distress.

Caterpillar has kept the large-engine portion of the company to this day but sold off the small-engine (aircraft) portion to Sundstrand Aviation of Rockford, Illinois. Sundstrand then merged with Hamilton Standard to become Hamilton-Sundstrand, a unit of United Technologies Corporation. Derivatives of the original Solar Titan gas turbine are now APUs for most of America's helicopter fleet, including *Marine 1*, the presidential helicopter. Another version of Solar's gas turbine serves as the starter engine (pony) for the Air Force F-16 fighter.

COMPARISONS

Make/Model	CID	Cyl.	Belt/PTO HP	Drawbar Pull (hp)	Fuel Cons.*	Weight (lb.)
Farmall 656 Hydro	**281**	**6**	**66.06**	**50.5**	**12.21**	**10,010**
Farmall 656D	**281**	**6**	**61.52**	**52.98**	**14.38**	**9,235**
Farmall 656G	**263**	**6**	**63.85**	**55.23**	**12.26**	**9,165**
Farmall 656G Hydro	**263**	**6**	**65.8**	**50.09**	**10.41**	**9,995**
Massey 180D	236	4	63.68	53.84	16.76	8,665

*horsepower hours per gallon
Weight in pounds is as tested including ballast (if any) and driver.

1971 Farmall 656
The 656 replaced the 706 in 1967, and production continued through 1972. The new tubular bar grille was the most prominent difference between the two models. This 656 is powered by gasoline, but the tractor was also available in LPG or diesel versions. International Harvester also built a high-clearance version of the 656.

The Farmall 756

When Harvester developed the 756 Farmall, they basically restyled the Model 706, bringing it up to date with the others in the 56 Series. The 756 was available in row-crop and high-clearance configurations. Because of its similarity to the 706, the University of Nebraska did not field test the 756.

In 1969, John Deere introduced the 4000, a bare-bones version of its popular 4020, which sold for about $2,000 less. To counter this new threat, International Harvester introduced the no-frills versions of the Farmall 756 and 856, renamed the 756 Custom and 856 Custom. These models featured smaller fuel tanks, single headlights, and single remote hydraulic outlets.

The Farmall 856

The Farmall 856 replaced the famous 806 in 1967. Although the 856 was restyled to bring it up to date and was beefed up with a new, larger-displacement six-cylinder engine of 406.9 ci, the tractor was very similar to its predecessor. The new diesel gave

1971 Farmall 856
International began building the 856 as a replacement for the venerable 806 in 1967. Harvester updated the tractor's styling and increased its engine displacement from 361 ci to 407 ci. Only the diesel version was available, and the increased displacement pushed the horsepower over 100 without the complications of a turbocharger. Tim Johnson of Atlanta, Indiana, added an M&W turbo to this tractor, boosting its horsepower to 169. International Harvester manufactured the 856 until 1971.

100 horsepower without the expense or complexity of a turbocharger. The 856 was available in high-clearance and regular row-crop versions. Although narrow- or tricycle-front row-crop tractors were still in the IH catalog, they were becoming a thing of the past on tractors as big as the 856. Gasoline and LPG engines from the 806 were carried in the catalog, but it is not known if any were sold. Only the diesel was tested at Nebraska.

COMPARISONS

Make/Model	CID	Cyl.	Belt/PTO HP	Drawbar Pull (hp)	Fuel Cons.*	Weight (lb.)
Farmall 856D	**407**	**6**	**100.49**	**87.17**	**15.15**	**12,035**
Ford 8000D	401	6	105.74	86.72	16.06	14,785

*horsepower hours per gallon
Weight in pounds is as tested including ballast (if any) and driver.

The Farmall 1256

IH replaced the Model 1206 with the Farmall 1256 in 1967. The new 1256 featured a turbocharged version of the 406.9-ci engine with a maximum PTO horsepower exceeding 116. Harvester engineers used a Schwitzer turbo in place of the Solar unit on the 1256 since by then the Solar Company had been sold to Caterpillar. The Farmall 1256 was available in row-crop configuration only.

1968 Farmall 1256
Successor to the 1206, the 1256 had the same displacement, but with a Schwitzer turbocharger instead of the Solar unit. In the Nebraska test, the 1256's horsepower was up to 116. The 56 Series Farmall featured improved operator convenience and comfort. For example, the 1256 offered a tilt steering wheel and a high-back chair with folding arms.

1969 Farmall 1256
Factory cabs were becoming the norm by the late 1960s. The cab delivered with the 1256 was known as the "Ice Cream Box." Dual rear tires were also very common by then, but the tricycle front end had all but disappeared.

COMPARISONS

Make/Model	CID	Cyl.	Belt/PTO HP	Drawbar Pull (hp)	Fuel Cons.*	Weight (lb.)
Farmall 1256TD	**407**	**6**	**116.12**	**102.03**	**15.55**	**13,810**
Oliver 2050D	478	6	118.78	96.36	15.33	18,890

*horsepower hours per gallon
Weight in pounds is as tested including ballast (if any) and driver.

SPECIFICATIONS

Model	656 (HD)	656 (D)	656 (HG)	656 (G)	856 (D)	1256 TD
Years	1966-1972	1966-1972	1966-1972	1966-1972	1967-1971	1967-1968
Test No.	967	912	968	909	970	971
Test Date	Nov. 1967	Sep. 1965	Nov. 1967	Sep. 1965	Apr. 1968	Apr. 1968
Engine Cyl./CID	6/281	6/281	6/263	6/263	6/407	6/407
Rated RPM	2,300	1,800	2,300	1,800	2,400	2,400
Drawbar HP	50.5	52.98	50.09	55.23	87.17	102.03
Belt/PTO HP	66.06	61.52	65.8	63.85	101.49	116.12
Drawbar Pull (lb.)	7,558	7,500	7,587	7,486	9,417	11,104
Test Weight (lb.)	10,010	9,235	9,995	9,165	12,035	13,810
Fuel	Diesel	Diesel	Gasoline	Gasoline	Diesel	Diesel
Transmission	Hydro	10f-2r	Hydro	10f-2r	16f-4r	16f-4r
Test Configuration	RC-Tri	RC-Tri	RC-Tri	RC-Tri	RC-WF	Standard
Top Road Speed (mph)	21	16	21	16	20	19
Price, New ($)	8,100	7,300	7,300	6,500	10,100	11,000
Production	35,000				25,000	3,000

The Farmall 544

The Farmall 544 was the second IH tractor to offer the optional hydrostatic transmission. This time it came with a dynamically balanced four-cylinder gasoline or diesel engine. The standard transmission was the traditional five-speed with Torque Amplifier. The gasoline version displaced 200 ci, while the diesel displaced 239 ci. Both produced a nominal 54 horsepower at the PTO.

The 544 was available in tricycle or adjustable wide-front configurations. A sliding gear two-range auxiliary gearbox was used with the hydrostatic transmission, giving speeds of up to 21 mph.

COMPARISONS

Make/Model	CID	Cyl.	Belt/PTO HP	Drawbar Pull (hp)	Fuel Cons.*	Weight (lb.)
Farmall 544G	**200.3**	**4**	**52.84**	**44.74**	**11.97**	**8,310**
Farmall 544D	**238.6**	**4**	**52.95**	**44.56**	**15.98**	**8,470**
Farmall 544G Hydro	**200.3**	**4**	**53.87**	**40.33**	**9.71**	**9,710**
John Deere 2520G	202.7	4	56.98	48.09	10.3	8,990
Massey MF-165G	212	4	51.91	44.44	10.85	7,760

*horsepower hours per gallon
Weight in pounds is as tested including ballast (if any) and driver.

The Farmall 1456

In 1969, the Farmall Model 1456 replaced the Model 1256. Even though it looked much the same, the 1456 was new from the bumper to the drawbar. The 1456, which featured an improved six-cylinder dry-sleeve DT407 turbo-diesel 407-ci engine, topped out at 131.8 PTO horsepower in Nebraska Test No. 1048. The new, optional two-door cab allowed entry from either side of the tractor. The 156 also featured rear brakes with 11-inch discs and a 3.5-inch diameter rear axle.

1970 Farmall 1456
Introduced in late 1969, the 1456 replaced the 1256, and although similar in appearance, it was considerably revised. The 1456 featured a new, dry-sleeve 407-ci turbocharged diesel engine with 132 horsepower. International also offered a new "Custom Cab" with a door on either side.

Despite its large size and short production run, the Farmall 1456 was quite popular. Between 1969 and 1971, IH produced almost 6,000 models.

COMPARISONS

Make/Model	CID	Cyl.	Belt/PTO HP	Drawbar Pull (hp)	Fuel Cons.*	Weight (lb.)
Farmall 1456D	**407**	**6**	**131.8**	**114.69**	**15.81**	**17,350**
John Deere 4320D	404	6	116.55	101.71	14.65	14,380

*horsepower hours per gallon
Weight in pounds is as tested including ballast (if any) and driver.

The Farmall 826

Available in either an eight-speed manual transmission with Torque Amplifier or the fully hydrostatic transmission, the Farmall 826 was about a 90-horsepower row-crop tractor. IH produced the 826 in both diesel- and gasoline-powered versions. The German-built 826 featured a six-cylinder diesel engine, which displaced 358 ci. The gasoline model, built in America, used the C-301 engine from the 806.

A painting of an International Hydro from a 1971 International Harvester calendar.

1970 Farmall 826 Hydro
The 826 tractor was built to fill the 90-horsepower gap between the 75-horsepower 700 series and the over-100-horsepower niche of the 800 series. It used the German-built 358-ci diesel. A gasoline version using the 301-ci engine from the 806 was also available. This particular example features the hydrostatic drive, but a conventional transmission was an option. Power brakes, power steering, and a Category II three-point hitch were standard. Harvester produced the 826 from 1969 through 1971. This 826 has extra headlights on the front and an M&W turbocharger, which boosts the horsepower to 104. Although this tractor had logged 8,200 hours in the field when the photo was taken, the hydro drive "had never been touched."

COMPARISONS

Make/Model	CID	Cyl.	Belt/PTO HP	Drawbar Pull (hp)	Fuel Cons.*	Weight (lb.)
Farmall 826D	**358**	**6**	**92.19**	**78.92**	**14.88**	**13,840**
Farmall 826D Hydro	**358**	**6**	**84.66**	**61.39**	**13.53**	**13,840**
Case 970D	401	6	85.31	71.56	15.71	13,870
Oliver 1855D	310	6	98.6	82.65	15.84	14,800

*horsepower hours per gallon
Weight in pounds is as tested including ballast (if any) and driver.

The Farmall 1026

The Farmall 1026 was the first tractor with more than 100 horsepower to be equipped with a hydrostatic drive, and one of the first Farmall models to feature the improved cab built by Excel Industries. Otherwise, the 1026 was quite similar to the 1256, with the same turbocharged DT407 six engine. The Torque Amplifier was not available on this model. Since only 2,414 of the Farmall 1026 model were delivered, the tractor is quite collectible today.

COMPARISONS

Make/Model	CID	Cyl.	Belt/PTO HP	Drawbar Pull (hp)	Fuel Cons.*	Weight (lb.)
Farmall 1026D	**407**	**6**	**112.45**	**80.24**	**14.3**	**14,975**
Oliver 1955D	310	6	108.16	90.32	15.78	16,220

*horsepower hours per gallon
Weight in pounds is as tested including ballast (if any) and driver.

The Farmall 766

Unveiled in the fall of 1971, the Farmall 766 was the smallest of the new 66 Series tractors that replaced the 56 Series. The 766 was available in either a 291-ci gasoline

SPECIFICATIONS

Model	544 (G)	544 (HG)	544 (D)	1456 TD	826 (D)	826 (HD)	1026 (HD)
Years	1968-1973	1968-1973	1968-1973	1969-1971	1969-1971	1969-1971	1970-1971
Test No.	984	1007	983	1048	1045	1046	1047
Test Date	Sep. 1968	May 1969	Sep. 1968	June 1970	June 1970	June 1970	June 1970
Engine Cyl./CID	4/200	4/200	4/239	6/407	6/358	6/358	6/407
Rated RPM	2,200	2,400	2,200	2,400	2,400	2,400	2,400
Drawbar HP	44.74	40.33	44.56	114.69	78.92	61.39	80.24
Belt/PTO HP	52.84	53.87	52.95	131.8	92.19	84.66	112.45
Drawbar Pull (lb.)	6,521	6,518	6,528	12,558	9,631	9,780	9,583
Test Weight (lb.)	8,310	9,710	8,470	17,350	13,840	13,840	14,975
Fuel	Gasolne	Gasoline	Diesel	Diesel	Diesel	Diesel	Diesel
Transmission	10f-2r	Hydro	10f-2r	16f-8r	16f-8r	Hydro	Hydro
Test Configuration	RC-WF	RC-WF	RC-WF	Standard	RC-Tri	RC-Tri	Standard
Top Road Speed (mph)	18	22	18	20	18	18	17
Price, New ($)	6,000	6,600	6,500	12,300	9,500	10,500	11,800
Production	———	5,500	———	5,600	—— 15,000	——	4,000

or 360-ci diesel version, both producing about 83 PTO horsepower. IH did make smaller tractors for the U.S. market, but they were in the utility configuration and were not Farmalls.

With the 66 Series tractors, IH replaced the "idiot lights" that filled the instrument panels of the previous series with real instruments—a move that farmers really appreciated.

Like others in the series, the 766 was based on the 756 it replaced but restyled and extensively beefed up, especially in the front end and frame. This was necessary because of the practice of carrying nitrogen application tanks on each side of the tractor's nose.

1974 Farmall 766 (and next page)
International Harvester produced gasoline- and diesel-powered 766s from 1971 to 1975. This 766 features an aftermarket Schwitzer turbo with an M&W kit, boosting the horsepower to 150. This 1974 model was originally equipped with a cab, but the owner traded it for a set of deluxe fenders.

The 766 used the standard eight-speed forward, four-speed reverse transmission. Options included power-shift, T/A, Front Wheel Assist (FWA), factory cab, and the new electrically actuated differential lock.

COMPARISONS

Make/Model	CID	Cyl.	Belt/PTO HP	Drawbar Pull (hp)	Fuel Cons.*	Weight (lb.)
Farmall 766G	**290.8**	**6**	**79.73**	**71.32**	**11.29**	**11,880**
Farmall 766D	**360**	**6**	**85.45**	**73.37**	**13.87**	**12,720**
Ford 7000D	256	4	83.49	66.66	16.47	10,795

*horsepower hours per gallon
Weight in pounds is as tested including ballast (if any) and driver.

The Farmall 966

The Farmall 966, which replaced the 826, operated at about 93 horsepower. The 966 was available in 414-ci six-cylinder wet-sleeve naturally aspirated diesel only. This tractor was similar to the Model 1066, but it did not come with a turbocharger.

The 966 came with either the hydrostatic drive or the standard eight-speed forward, four-speed reverse transmission with the optional power-shift T/A. Options included FWA, a new improved Deluxe Cab, the less-fancy Custom Cab, and the two-post ROPS and shade. Farmers could order the Farmall 966 in the high-clearance configuration. The dual-shaft/two-speed PTO and the differential lock were available.

COMPARISONS

Make/Model	CID	Cyl.	Belt/PTO HP	Drawbar Pull (hp)	Fuel Cons.*	Weight (lb.)
Farmall 966D	**414**	**6**	**96.01**	**80.4**	**14.85**	**13,500**
Farmall 966D Hydro	**414**	**6**	**91.38**	**68.08**	**12.76**	**12,875**
Case 970D	401	6	93.41	79.85	15.05	14,510

*horsepower hours per gallon
Weight in pounds is as tested including ballast (if any) and driver.

1975 Farmall 966
Built from 1971 to 1975, the 966 was available with a diesel engine only. Its 414-ci six-cylinder engine had about 100 horsepower. Standard equipment included a differential lock.

The Farmall 1066

The Farmall 1066, featuring the turbocharged version of the 414-ci six-cylinder wet-sleeve diesel, offered a nominal 115 horsepower. The 1066 came with either the standard eight-speed forward, four-speed reverse transmission with the optional power-shift T/A or the infinitely variable hydrostatic drive. Options included FWA, a new improved Deluxe Cab, the two-post ROPS with or without the shade, the dual-shaft/two-speed PTO, and the differential lock. IH also produced a high-clearance configured Farmall 1066.

COMPARISONS

Make/Model	CID	Cyl.	Belt/PTO HP	Drawbar Pull (hp)	Fuel Cons.*	Weight (lb.)
Farmall 1066D	**414**	**6**	**116.23**	**100.5**	**15.16**	**15,170**
Farmall 1066D Hydro	**414**	**6**	**113.58**	**84.88**	**14.32**	**13,190**
Case 1070D	451	6	107.36	90.77	15.10	17,520

*horsepower hours per gallon
Weight in pounds is as tested including ballast (if any) and driver.

1975 Farmall 1066 Turbo
A big tractor in most circles, the 1066 weighed 7 tons in working trim. The 1066's 414-ci wet-sleeve six-cylinder turbocharged diesel belted out more than 115 horsepower. Like all tractors in the 66 series, the 1066 had remote hydraulic couplings that could be coupled and decoupled under pressure without the loss of fluid. This tractor is all-gear drive, but a "Hydro" version was also available.

AND SON
Turbo
INTERNATIONAL
1066
FARMALL

SPECIFICATIONS

Model	766 (D)	766 (G)	966 (D)	966 (HD)	1066 TD	1066 (H) TD
Years	1971-1975	1971-1975	1971-1975	1971-1975	1971-1976	1971-1976
Test No.	1117	1094	1082	1095	1081	1083
Test Date	Dec. 1972	May 1972	Oct. 1971	May 1972	Oct. 1971	Oct. 1971
Engine Cyl./CID	6/360	6/291	6/414	6/414	6/414	6/414
Rated RPM	2,600	2,400	2,400	2,400	2,400	2,400
Drawbar HP	75.37	71.32	80.4	68.08	100.5	84.88
Belt/PTO HP	85.45	79.73	96.01	91.38	116.23	113.58
Drawbar Pull (lb.)	9,236	8,841	10,041	8,909	11,625	9,576
Test Weight (lb.)	12,720	11,880	13,500	12,875	15,170	13,190
Fuel	Diesel	Gasoline	Diesel	Diesel	Diesel	Diesel
Transmission	16f-8r	16f-8r	16f-8r	Hydro	16f-8r	Hydro
Test Configuration	Standard	RC-Tri	Standard	Standard	Standard	Standard
Top Road Speed (mph)	20	20	18	18	20	17
Price, New ($)	9,700	8,800	11,600	12,700	12,900	13,900
Production	—— 11,000 ——		—— 25,000 ——		—— 55,000 ——	

The Farmall 1466

The Farmall 1466 replaced the 1456 in 1971. It featured the beefed-up front end of the 66 Series. The 1466 was not available in a high-clearance version or with the differential lock available on the 1456. The Deluxe Cab, with improved visibility and soundproofing, was an option. Only the 1,000-rpm PTO was offered. On 1973 models of the 1466 (and others in the 66 Series), IH replaced the optional two-post ROPS (with optional shade) with an improved four-post design. The ROPS was standard on tractors delivered without cabs.

The Farmall 1466 featured a new robust turbocharged DT-436 (436 ci), which produced a nominal 145 horsepower at the PTO. The eight-speed transmission was available with the Torque Amplifier.

International Harvester officially dropped the Farmall name in 1973, but some Farmall tractors that were in production before 1973 carried the Farmall badge until 1975.

1972 Farmall 1466 Turbo
With a 436-ci six-cylinder turbocharged diesel engine of 145 horsepower, the 1466 was near the high end of farm power during its production years, 1971 to 1975.

COMPARISONS

Make/Model	CID	Cyl.	Belt/PTO HP	Drawbar Pull (hp)	Fuel Cons.*	Weight (lb.)
Farmall 1466D	**436**	**6**	**145.77**	**123.2**	**15.19**	**16,600**
Ford 9600DT	401	6	135.46	111.4	16.09	18,000

*horsepower hours per gallon
Weight in pounds is as tested including ballast (if any) and driver.

The Farmall 666

Bravely ignoring the possibility of criticism from religion fundamentalists (666 is the number of the Satanic Beast of Revelation 13:18), IH introduced the Farmall 666 in 1972—a year after the others in the 66 Series. Its predecessor, the Farmall 656, was selling so well that management didn't want to interrupt production, so they delayed the release of the 666.

The Farmall 666 was available with either a gasoline or diesel six-cylinder engine. While the diesel version displaced 312 ci and the gasoline version displaced 291 ci, both produced around 70 PTO horsepower. The 666 featured a five-speed transmission with the Torque Amplifier.

Although the 666 came with most of the features offered on the larger 66 Series tractors, this model was only available with a 540-rpm PTO.

COMPARISONS

Make/Model	CID	Cyl.	Belt/PTO HP	Drawbar Pull (hp)	Fuel Cons.*	Weight (lb.)
Farmall 666G	**290.8**	**6**	**66.3**	**57.29**	**11.95**	**10,050**
Farmall 666D	**312**	**6**	**66.29**	**58.01**	**14.23**	**10,330**
Massey MF-265D	236	4	60.73	50.38	15.9	8,200
Massey MF-275D	248	4	67.43	60.73	14.64	9,330

*horsepower hours per gallon
Weight in pounds is as tested including ballast (if any) and driver.

The Farmall 1468

In 1967, rival Massey-Ferguson introduced its Model 1150, which was powered by a 511-ci 135-horsepower Perkins V-8 diesel engine. Massey had acquired the British Perkins engine company in 1959. True to form, the IH marketing gurus insisted that Harvester have a V-8 too, so the company introduced the Farmall 1468 in 1971. Aside from the 1468's V-8 diesel engine, the tractor was essentially the same as the 1466. Known as the DV-550 (550 ci), the big, naturally aspirated V-8 produced a PTO horsepower of 145.

An interesting feature of the DV-550 was its ability to run on only four of its eight cylinders when at idle, or low loads. General Motors used this same feature on its mid-1970s Cadillacs as a fuel-saving measure. In the case of the DV-550, cylinders one, four, six, and seven only received fuel under conditions of light load, and the valve lifters on the other cylinders were vented to prevent the valves from opening.

COMPARISONS

Make/Model	CID	Cyl.	Belt/PTO HP	Drawbar Pull (hp)	Fuel Cons.*	Weight (lb.)
Farmall 1468D	**550**	**8**	**145.49**	**127.70**	**15.60**	**16,310**
Massey MF-1155D	540	8	140.97	117.80	15.57	20,460
Minn-Moline G-1355D	585	6	142.66	125.66	15.21	18,070
Oliver 2255D	573	8	142.62	123.72	15.59	17,990

*horsepower hours per gallon
Weight in pounds is as tested including ballast (if any) and driver.

(this and previous page)

1972 Farmall 1468

Harvester built some 2,900 of these V-8 diesel tractors between 1971 and 1974. The tractor was essentially the same as the Farmall 1466 but outfitted with a 550-ci engine from the IH truck division. Under light or no load, the injectors fire only four of the 1468's eight cylinders. As the load increases, the remaining cylinders fire. The twin straight pipes on this 1468 give the tractor a touch of class.

1974 Farmall 1468
Harvester's 550-ci V-8 diesel is evident from the twin exhaust stacks. The engine had a bore and stroke of 4.50x4.3125 inches and a rated speed of 2,600 rpm. The tractor's working weight was about 16,000 pounds.

The Farmall Hydro 100 and Hydro 70

Harvester dealers were besieged with complaints that hydrostatic-drive tractors could not keep pace with gear-drive tractors of the same model number. Patiently explaining that such performance losses were the price one paid for the flexibility did not, as they say, feed the bulldog. The farmers assumed that because they had paid a premium for the hydrostatic tractor, they should not be shown up in the field. Harvester could not increase the horsepower of the hydrostatic models because the hydraulic units were already at their limits in most cases. In an attempt to remedy the problem at the lowest cost, IH simply rebadged the 966 and 1066 Farmalls as Hydro 100s. This subterfuge only alienated farmers who had already had hydrostatic-drive tractors. In the new models, IH replaced the 414-ci engine used in these two tractors with the naturally aspirated 436-ci six. The company then retagged the Farmall 666 as the Hydro 70 without making any other changes to the design. A new white stripe over the hood identified these models.

COMPARISONS

Make/Model	CID	Cyl.	Belt/PTO HP	Drawbar Pull (hp)	Fuel Cons.*	Weight (lb.)
Farmall Hydro 100D	**436**	**6**	**105.02**	**81.59**	**12.92**	**13,430**
Ford 8700D	401	6	110.58	97.96	14.29	12,270

*horsepower hours per gallon
Weight in pounds is as tested including ballast (if any) and driver.

1976 Farmall Hydro 70
Made from 1973 to 1976, the Hydro 70 was simply a renamed version of the Hydro 666. Harvester renamed the 666 to avoid direct performance comparisons between the gear drive and hydro drive versions. The unavoidable loss of power in the hydrostatic drive was the price paid for the additional flexibility it offered. This one has had a black stripe added by the owner.

The Farmall 1566

Locked in a horsepower race with the other major U.S. tractor suppliers, IH updated two of its biggest tractors, placing them into the over-150-horsepower class. The first was the 1566, the model that replaced the 1466. Because this tractor was produced as IH was in the process of phasing out the Farmall line, some of the 1566s sported the Farmall tag while others did not. To avoid overstressed final-drive problems that had proven so costly on the 460 and 560 models, engineers designed a new planetary final drive. They also boosted the six-cylinder 436-ci turbocharged DT-436 to 160 horsepower by increasing the turbocharger pressure and the injector stroke. The 1566 featured twelve forward speeds and the partial-range power-shift Torque Amplifier.

COMPARISONS

Make/Model	CID	Cyl.	Belt/PTO HP	Drawbar Pull (hp)	Fuel Cons.*	Weight (lb.)
Farmall 1566DT	**436**	**6**	**161.01**	**140.35**	**14.55**	**18,260**
White 2-155DT	478	6	157.73	133.59	15.12	16,650

*horsepower hours per gallon
Weight in pounds is as tested including ballast (if any) and driver.

The Farmall 1568

The second of the revised "horsepower-race" tractors was the 1568, which replaced the 1468. At least some of these tractors wore the Farmall tag. IH engineers upgraded the naturally aspirated DV-550 engine of the 1568 to produce just over 150 horsepower by making changes to the head and piston. The 1568 was much the same as the 1566 tractor, with a few minor changes—the cab was standard equipment and the transmission was a twelve-speed unit with the T/A standard.

COMPARISONS

Make/Model	CID	Cyl.	Belt/PTO HP	Drawbar Pull (hp)	Fuel Cons.*	Weight (lb.)
Farmall 1568DT	**550**	**8**	**150.70**	**133.79**	**14.94**	**14,950**
Allis-Chalmers 7040DT	426	6	136.49	117.12	15.19	11,795
Allis-Chalmers 7060DTI	426	6	161.51	140.08	15.87	13,990

*horsepower hours per gallon
Weight in pounds is as tested including ballast (if any) and driver.

SPECIFICATIONS

Model	1466 TD	666 (G)	666 (D)	1468 (D)	Hydro 100 (D)	1566 TD	1568 (D)
Years	1971-1976	1972-1975	1972-1975	1971-1974	1973-1976	1974-1976	1974-1975
Test No.	1080	1152	1151	1118	1158	1174	1175
Test Date	Oct. 1971	Dec. 1973	Nov. 1973	Dec. 1972	May 1974	May 1975	May 1975
Engine Cyl./CID	6/436	6/291	6/312	8/550	6/436	6/436	8/550
Rated RPM	2,400	2,000	2,000	2,600	2,400	2,600	2,600
Drawbar HP	123.2	57.29	58.01	127.7	81.59	140.75	133.79
Belt/PTO HP	145.77	66.3	66.29	145.49	105.02	161.01	150.7
Drawbar Pull (lb.)	13,165	7,981	8,012	12,932	9,801	15,774	14,918
Test Weight (lb.)	16,660	10,050	10,330	16,310	13,430	18,260	17,350
Fuel	Diesel	Gasoline	Diesel	Diesel	Diesel	Diesel	Diesel
Transmission	16f-8r	10f-2r	10f-2r	16f-8r	Hydro	12f-6r	12f-6r
Test Configuration	Standard	Standard	Standard	Standard	Standard	Standard	Standard
Top Road Speed (mph)	20	17	17	22	17	18	18
Price, New ($)	14,800	7,600	8,500	18,000	14,000	23,000	23,000
Production	22,000	—— 6,000 ——		2,500	7,000	9,000	1,000

The End

Harvester's downhill slide began in 1958 with the Farmall 460/560 final drive debacle. Their impeccable reputation in farm tractors was tarnished. Even with the service people's sterling performance in replacing the broken parts under extended warranties, the damage was done. Frank Jenks replaced John McCaffrey as CEO in 1958 and ordered the engineers to make the next series of tractors "hell for stout, or else!" They did!

But shortly thereafter, in 1960, John Deere introduced its "New Generation" tractors. The timing couldn't have been worse for Harvester. As far back as 1953, John Deere management had commissioned a secret team of engineers in a secret location to come up with a series of "clean-sheet-of-paper" tractors. These tractors were so advanced that they were virtually unchanged until after the Harvester Ag Group was absorbed into Case in 1985.

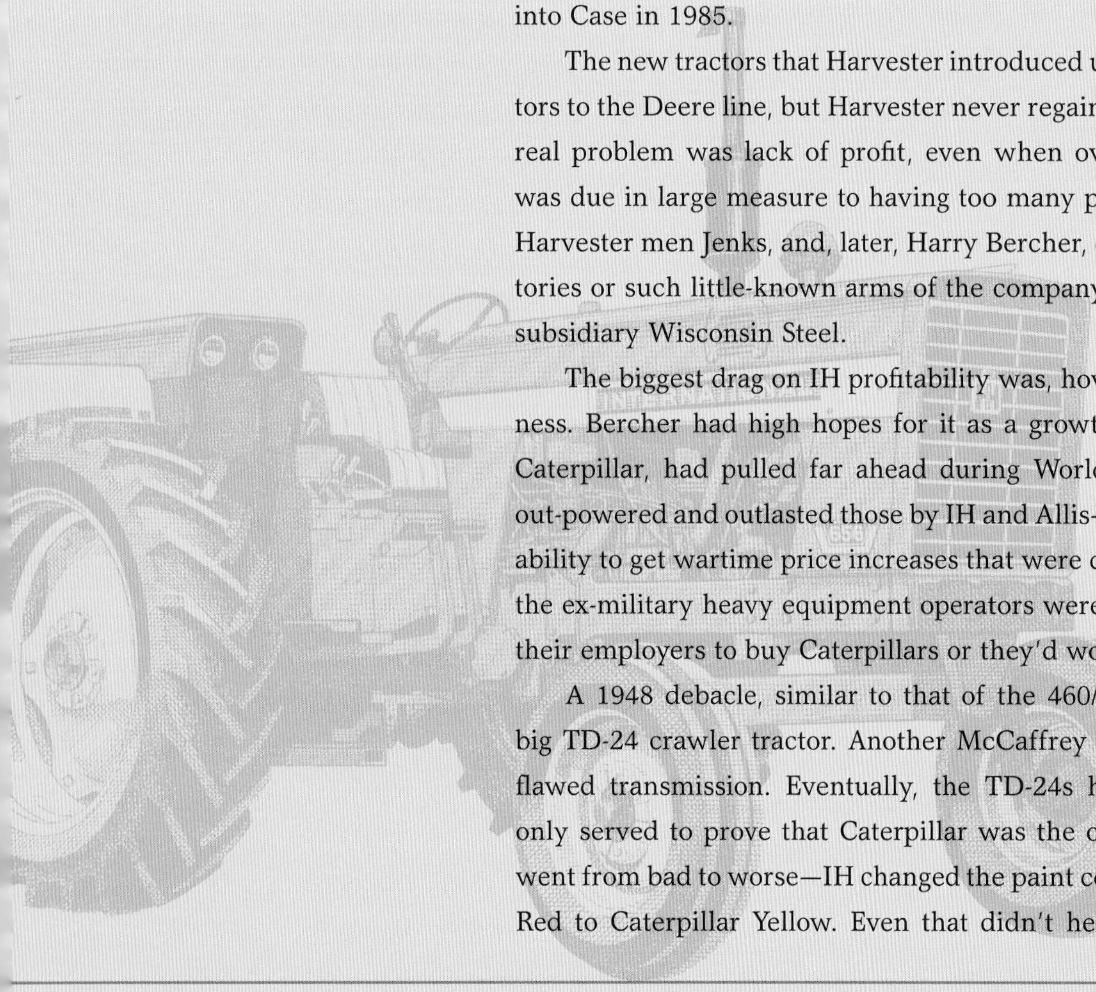

The new tractors that Harvester introduced under Frank Jenks were worthy competitors to the Deere line, but Harvester never regained its previous momentum. Harvester's real problem was lack of profit, even when overall sales were up in the 1960s. This was due in large measure to having too many products made in obsolete facilities. Old Harvester men Jenks, and, later, Harry Bercher, couldn't bring themselves to axe old factories or such little-known arms of the company as the wholly owned and unprofitable subsidiary Wisconsin Steel.

The biggest drag on IH profitability was, however, the construction machinery business. Bercher had high hopes for it as a growth industry. Their principal competitor, Caterpillar, had pulled far ahead during World War II with excellent machines that out-powered and outlasted those by IH and Allis-Chalmers. Further, Cat had the uncanny ability to get wartime price increases that were denied the others. After the war, most of the ex-military heavy equipment operators were experienced on Cat machines and told their employers to buy Caterpillars or they'd work for the ones that did!

A 1948 debacle, similar to that of the 460/560 rear-end problems, arose with the big TD-24 crawler tractor. Another McCaffrey rush-to-production resulted in a fatally flawed transmission. Eventually, the TD-24s had to be recalled and scrapped. This only served to prove that Caterpillar was the one choice in heavy equipment. Things went from bad to worse—IH changed the paint color of its construction line from Farmall Red to Caterpillar Yellow. Even that didn't help. Eventually, IH was able to free up some cash by selling off its Hough operations and its construction equipment lines to Dresser Industries.

In the late 1960s, the inflation caused by the war in Vietnam aggravated Harvester's financial woes. Labor and material costs escalated, exacerbating the drain of inefficient old factory buildings.

Brooks McCormick brought in Archie McCardell from Xerox as CEO to replace the retiring Bercher in 1977. McCardell had a "gung ho" reputation as a corporate savior. He soon found that cash-strapped IH could not afford the kind of long-range planning and research underway at Deere and Caterpillar. McCardell really met his Waterloo (no pun intended) when he sought concessions from the United Automobile Workers (UAW) in 1979. He was after holding the line on wages and benefits but also relief from the maze of archaic work rules that had developed over time. The workers reacted as though he was trying to break the union!

When McCardell had come aboard in 1977, he had been promised a huge bonus when a certain performance formula was met. In 1979, in an effort to avoid shortages in the event of a labor strike, the various business groups had increased production to boost inventory of both parts and finished products. With this surge in production, elements fell into place and McCardell was awarded a $1.8 million bonus. Needless to say, the UAW did not see that as proper when they had been asked to yield on wages and benefits.

The result was a devastating six-month strike that shattered employee morale and cost the company almost $600 million. Worse, the Federal Reserve had raised the prime interest rate to 20 percent, a move that drastically affected sales for Harvester and other implement manufacturers. Farmers and truckers could not buy equipment at such a high interest rate and still make money. The Carter administration's grain embargo of 1980, forbidding the shipment of grain to the Soviet Union, further hurt the farmer. Harvester's sales for 1980 dropped by almost one-third. Harvester was stuck with aging inventories.

In 1982, Archie McCardell resigned, knowing that his best efforts for the company were failing. Donald D. Lennox was made president and CEO. By then, some great new two- and four-wheel-drive tractors were on the market, with the new Synchrotorque transmission and improved diesel engines. High interest rates and a drought in 1983 kept sales sluggish. The inventories built up during the strike burdened dealers and hurt sales of the new tractors. Harvester was now fighting for its life and losing. Lennox made drastic cuts, reducing Harvester's far-flung empire to just three businesses: agricultural, trucks, and engines.

The picture for the farmer and for the "Ag" portion of International Harvester did not improve in 1983 or 1984. Industry-wide sales of large tractors continued to drop to about 35 percent of 1979 levels. In November 1984, Lennox accepted Tenneco's offer to buy the Harvester Agricultural Equipment Group. Tenneco was a Houston-based conglomerate with a wholly owned subsidiary—another old-line pioneer in the tractor business—J. I. Case. International Harvester was no more.

Although the merger made for strange bedfellows, some outstanding tractors have emerged. At the end of the twentieth century, Case-IH was folded into the New Holland conglomerate principally owned by Fiat. Nevertheless, International Harvester, and especially the Farmall tractor, will long be remembered for major contributions to world agriculture.

ACKNOWLEDGMENTS

The author and photographer are indebted to many people who helped make this book possible. The following Farmall collectors not only gave us days of their valuable time, provided tractors to photograph, and helped with necessary technical and historical facts, they sometimes fed us, washed tractors for us, introduced us to neighboring collectors, and tolerated schedule mix-ups with grace. In their dedication to the hobby and for just being all-around nice people, Farmall collectors are second-to-none.

We thank Clete Ahler of Burlington, Wisconsin; Dick Alberts of Grand Rapids, Michigan; Louis Bunker of Lena, Illinois; John Duitsman of Gifford, Illinois; Wilson Gatewood of Noblesville, Indiana; Darius Harms of St. Joseph, Illinois; Ronald Hattendorf of Genoa, Illinois; Tim Johnson of Atlanta, Indiana; Gary Keister of Patoka, Illinois; Dan and Mary Lou Langy of Lena, Illinois; Harry Lee of Elnora, Indiana; Buryl McFadden of Beaver Creek, Ohio; Alvin Mennenga of Thomasboro, Illinois; Neil Michel of Tipton, Indiana; Ronald Neese of Noblesville, Indiana; Jim Perrin of Harlan, Indiana; Bill Powell of Tipton, Indiana; Norval Poyser of Middlebury, Indiana; Greg Schmitt of Tipton, Indiana; Norman Sevcik of Northfield, Minnesota; Donald Shaefer of Grabill, Indiana; Marty Theime of Noblesville, Indiana; Bev and Dale Thompson of Genoa, Illinois; John Wagner of White Pigeon, Michigan; Donald Wolf of Ft. Wayne, Indiana; and Kevin and Rick Wolkin of Urbana, Illinois. We would also like to thank John Pritchard, Alan Davies, Richard Seale, Colin Shearn, and Tony Gerrard in Great Britain; Johann and Gerlinde Hood in Germany; and Sylvere Portier in France. Finally, thanks also to Dennis Pernu, Zack Miller, Cindy Laun, and the rest of the staff at Motorbooks for helping bring this new edition to fruition.

Andrew Morland
Robert N. Pripps

INDEX

Brimming with creative inspiration, how-to projects, and useful information to enrich your everyday life, quarto.com is a favorite destination for those pursuing their interests and passions.

First Published in this format in 2020 by Motorbooks, an imprint of The Quarto Group,
100 Cummings Center, Suite 265-D, Beverly, MA 01915, USA.
T (978) 282-9590 F (978) 283-2742 Quarto.com

Motorbooks titles are also available at discount for retail, wholesale, promotional, and bulk purchase.
For details, contact the Special Sales Manager by email at specialsales@quarto.com or by mail at The Quarto Group, Attn: Special Sales Manager, 100 Cummings Center, Suite 265-D, Beverly, MA 01915, USA.

24 23 22 2 3 4 5

ISBN: 978-0-7603-6389-8

Library of Congress Cataloging-in-Publication Data is available.

Acquiring Editor:
Dennis Pernu

Art Director:
Cindy Samargia Laun

Cover and Page Design:
Brad Norr Design

Photography:
All modern color photography
of tractors (including front cover) by
Andrew Morland/Motorbooks Archive,
unless indicated otherwise.

Printed in China